Wheat Yellow Rust in the Extended Himalayan Regions and the Middle East

Editors
Mingju Li & Sajid Ali

China Agriculture Press
Beijing

Editors

Mingju Li

Yunnan Key Laboratory of Green Prevention and Control of Agricultural Transboundary Pests, Agricultural Environment and Resources Institute, Yunnan Academy of Agricultural Sciences, Kunming, China

Sajid Ali

Department of Agriculture (Plant Breeding and Genetics), Hazara University, Mansehra, Pakistan

Contributors

Emad Mahmood Ghaleb Al-Maaroof

College of Agricultural Engineering Sciences, University of Sulaimani, IKR, Iraq

Baidya Nath Mahto

Nepal Agricultural Research Council, Singhdarbarplaza, Kathmandu, Nepal

Ahamd Abbasi Moghaddam

Seed and Plant Improvement Institute, Agricultural Research, Education and Extension Organization, Karaj, Iran

Mohamed A. Gad

Plant Pathology Research Institute, Agricultural Research Center, Giza, Egypt

Haoxing Li

State Key Laboratory of Agricultural Microbiology, College of Life Science and Technology, Huazhong Agricultural University, Wuhan, China

Chi He

Yunnan Key Laboratory of Green Prevention and Control of Agricultural Transboundary Pests, Agricultural Environment and Resources Institute, Yunnan Academy of Agricultural Sciences, Kunming, China

Zahoor A. Swati

Institute of Biotechnology & Genetic Engineering, The University of Agriculture, Peshawar, Pakistan

Reda I. Omera

Plant Pathology Research Institute, Agricultural Research Center, Giza, Egypt

Muhammad Rameez Khan

Institute of Biotechnology & Genetic Engineering, The University of Agriculture, Peshawar, Pakistan

Atef A. Shahin

Plant Pathology Research Institute, Agricultural Research Center, Giza, Egypt

Aamir Iqbal

Institute of Biotechnology & Genetic Engineering, The University of Agriculture, Peshawar, Pakistan

Zia-ur-Rehman

Institute of Biotechnology & Genetic Engineering, The University of Agriculture, Peshawar, Pakistan

Muhammad Awais

Northwest Agriculture and Forestry University, Yangling, China

Ghulam Ullah

International Maize and Wheat Improvement Center (CIMMYT), Islamabad, Pakistan

Ehtisham Shakeel Khokhar

International Maize and Wheat Improvement Center (CIMMYT), Islamabad, Pakistan

Muhammad Imtiaz

International Maize and Wheat Improvement Center (CIMMYT), Islamabad, Pakistan

Muhammad Fayyaz

National Agriculture Research Station, Islamabad, Pakistan

Introduction to major contributors

Prof. Dr. Mingju Li

Yunnan Key Laboratory of Green Prevention and Control of Agricultural Transboundary Pests, Agricultural Environment and Resources Institute, Yunnan Academy of Agricultural Sciences, Kunming 650205, China

lily69618@163.com

Dr. Mingju Li, Professor of plant pathology at Agricultural Environment and Resources Institute, Yunnan Academy of Agricultural Sciences, Kunming, China. She mainly engaged in wheat rust research, focus on epidemiology, virulence characterization and molecular population genetics of pathogen, wheat varieties resistance assessment and resistance genes research, etc. Dr. Li got PhD in plant pathology from Chinese Academy of Agricultural Sciences in 2013, and has been studied as a visiting scholar at Washington State University, USA in 2016. She served as the Post Expert of the provincial Wheat and Barley Industrial System form 2020 to 2021. Since 2021, Dr. Li has been serving as the director of Wheat Pest Professional Committee of Plant Protection Society in Yunnan province, as well as the Principal Investigator of wheat diseases research group in her institute. Currently, Dr. Li serves as a reviewer for several international journals such as *Plant Disease, Microbiology Spectrum, European Journal of Plant Pathology*, etc. She has conducted several national projects, and has published more than 50 research articles so far, ten of them were cited by SCI, fifteen were included in Chinese core periodicals, and seven were collected in international conference proceedings. She has awarded both national and provincial Science and Technology Progress Awards.

Prof. (Associate) Dr. Sajid Ali

Department of Agriculture (Plant Breeding and Genetics), Hazara University, Mansehra, Pakistan

bioscientist122@yahoo.com

Dr. Sajid Ali, has acquired PhD in Genetics, Evolution and Population Biology from Paris-XI University, Paris, France. Dr. Ali has completed one Post-Doctorate in Population Genetics from Aarhus University, Denmark, and another from CIRAD-INRA-SupAgro University, Montpellier in crop and pathogen population genomics. He has expertise in crop and pathogen population dynamics at spatial and temporal scale along with in depth knowledge of genetics, breeding and ecological consequences of resistance genes deployment. Dr. Sajid Ali has a strong research group, initiated at University of Agriculture, Peshawar, and now at Hazara University, Mansehra, Pakistan, which is focused on understanding crop and pathogen diversity and population structure in the context of invasion and adaptation. The research work conducted in collaboration with other internationally recognized research groups has enabled to generate substantial results on wheat rust population biology with worldwide importance. Dr. Ali has earned several research grants from international donors like European Research Council-EU, CIMMYT-USDA-USA, JIC-UK and HEC-Pakistan. The work on crop improvement contributed to the development of wheat variety "Ghanimat-e-IBGE" , Brassica varieties "Dalai" and "Rokhana" and Common Bean variety "Himalaya-1" . The group has strong collaboration with research groups at international level (e.g., in Australia, Denmark, Canada, France, Mexico and UK), regional level (e.g., in China, Nepal, Bhutan, Iran and Afghanistan) and national level. He has research publication in internationally reputed journals like *PLoS Pathogens, Molecular Ecology, Molecular Ecology Resources, Plant Pathology, Plant Disease, Fungal Genetics and Biology, Phytopathology and Ecology and Evolution*. Dr. Ali has been invited to talk at various international conferences including events at Oxford University of UK, NIAB, Cambridge of UK, Northwest Agriculture and Forestry University of China and Aarhus University of Denmark. Dr. Ali has been awarded Research Productivity Award twice (in 2016 and 2017) and Fakhre-Peshawar Award (2019).

Prof. Dr. Emad Mahmood Ghaleb Al-Maaroof

Biotechnology and Crop Science Dept., College of Agricultural Engineering Sciences, University of Sulaimani, IKR, Iraq

emad.ghalib@univsul.edu.iq

Dr. Emad Mahmood Ghaleb Al-Maaroof, Professor of plant pathology and cereals diseases at the College of Agricultural Engineering Sciences at the University of Sulaimani, IKR, Iraq. He engaged in wheat rust research programs that focus on disease resistance and management, epidemiology, physiological specialization, virulence analysis and pathogen molecular characterization. He recently worked as a consultant at International Center for Agricultural Research in the Dry Areas (ICARDA) as Coordinator of the Harmonized Support for Agricultural Development in Iraq (HSAD) project, as well as the Head of the (ICARDA) Office in Erbil, IKR, to develop Iraq's agriculture sector. He was previously the Head of Plant Pathology Department at Iraq's Ministry of Science and Technology. He has developed over twenty new wheat and barley cultivars that are resistant to biotic and abiotic stresses. He has received three international scientific awards and six local awards in the field of food security and agricultural innovation from various international organizations, scientific centers, and universities. He has over 100 scientific journals publications and attended over 180 international and local conferences. He has served on the editorial boards of three international scientific journals and as a reviewer for over 40 local, regional and international scientific journals. He supervised a large number of postgraduate students, doctoral and master's degree holders, as well as holders of higher and initial degrees.

Prof. Dr. Baidya Nath Mahto
Nepal Agricultural Research Council, Singhdarbarplaza, Kathmandu, Nepal
bnmahto_7@hotmail.com

Dr. Baidya Nath Mahto involved in plant pathology research with special thrust to wheat pathology, rusts and foliar blights in particular. His research focus on wheat varietal resistance, virulence characterization, pathogenic variability, gene postulation, epidemiology, etc. He received PhD from Indian Agricultural Research Institute (IARI), New Delhi, India in 2000. He has been awarded with several honors and awards. He went for Fulbright visiting research scholar for a year during 2010 in North Dakota State University (NDSU), USA. He got Wheat Improvement Training at CIMMYT, Mexico for a year during 1991. He has visited several universities and research institutions as a visiting scientist abroad. He worked as an Executive Director and Principal Scientist (Plant Pathology) of Nepal Agricultural Research Council (NARC), Kathmandu, Nepal and served for more than 30 years and also hold different positions as Director of Planning and Co-ordination, Director of Financial Administration and Chicf of Plant Pathology, etc. He has been working as an adjunct professor at Himalayan College of Agricultural Sciences and Technology (HICAST) under Purbanchal University, Kathmandu, Nepal. Presently, he serves as Honorable Member, Provincial Policy and Planning Commission, Province No.2 Janakpurdham, Nepal and prepared Policies Plan especially for Agriculture sector and played significant role in preparation of Periodic Plan and Policies, monitoring and evaluation. He had conducted several projects and published more than 120 research articles in national and international journals and proceedings. He worked as a chair for Conservation Agriculture for Sustainable Intensification (CASI) Platform in South Asia 2018 and national expert and resource person for Pest Risk Analysis (PRA) in context of WTO.

Prof. (Assistant) Dr. Ahmad Abbasi Moghadam

Seed and Plant Improvement Institute (SPII), Agricultural Research, Education and Extension Organization (AREEO), Karaj, Iran

abasimoghadam@gmail.com

Dr. Ahmad Abbasi Moghadam is a member of the Scientific Board of SPII, AREEO. He investigates, evaluates, and characterizes plant germplasm accessions by classic and molecular techniques for seed health and disease resistance. He got his PhD in Plant Pathology from Indian Agricultural Research Institute, New Delhi, India, in 2003 with the scholarship award from AREEO, Iran. He was attend training programs on wheat rusts by ICARDA and later CIMMYT in Syria and Kenya. He has worked as a member of the Scientific Board of SPII, AREEO, head of seed health lab, deputy head of genetic department and national plant gene bank of Iran, SPII, member of Research Advisory Board, Head of Information technology and computer of SPII, member of the Boards of Trustees of the Iranian Mycological Society. He was supervisor/co-advisor of 14 MSc/PhD thesis project and published more than 75 research articles.

Prof. (Assistant) Dr. Mohamed A. Gad

Plant Pathology Research Institute, Agricultural Research Center, Giza 12619, Egypt

mohamedabo2002@yahoo.com

Dr. Mohamed A. Gad is a well-recognized scientist at the Plant Pathology Research Institute, Agricultural Research Center, Egypt and is a member in the Egyptian national campaign for wheat. He has worked on wheat and rust diseases with many good publications, which focus on disease resistance and epidemiology, management, physiological specialization, virulence analysis and pathogen molecular characterization. Dr. Gad has established a good national network related to his research work along with strong international linkages. He got his PhD in Biotechnology from Gauhati University, India in 2012. He has completed post-doctorate in Agricultural Environment and Resources Institute, Yunnan Academy of Agricultural Sciences, China in 2019. Dr. Gad is oriented towards the field and greenhouse testing of wheat germplasm along with tracking the pathogen population.

Foreword

Wheat (*Triticum aestivum* L.) is the most important food crop worldwide, which provides one-fifth of the protein and calories to more than 4.5 billion people. Yellow rust (also called stripe rust) caused by *Puccinia striiformis* f. sp. *tritici,* is one of the most destructive wheat diseases, which can result in significant yield losses or huge investment in chemical control. The extended Himalaya region (such as Nepal, Pakistan and China) is identified as a hotspot of sexual recombination and genetic diversity as well as the putative center of origin of the pathogen, while the Middle East is the center of origin for common wheat along with the source of high-temperature adapted strain. These areas are critical for wheat yellow rust epidemic around the world. Therefore, it is crucial to know the details about wheat yellow rust situation and research progress in these hotspot regions. However, there was a lack of such a comprehensive book regarding wheat yellow rust focusing on these important regions. This was an incentive for us to write this book.

Compilation of information on yellow rust research in these regions is further important due to the potential invasions across these areas and to other parts of the world, as reported in recent past. There is a dire need for collaborative efforts at the regional and global level to anticipate and adopt preemptive measures to avoid such invasions. To initiate such a collaboration, it is highly important to provide a comprehensive background of various research work done in different countries. This provides another reason to compile such a book with contribution from experts from various countries of these regions.

This book, covers the countries such as China, Nepal, Pakistan, Iran, Iraq, and Egypt. Each country is covered in one chapter, then the book includes six chapters, where each chapter addresses the research progress on different aspects of yellow rust. The chapters are arranged considering a rough order of the geographical position of the countries with a direction of east to west. The contents encompass disease epidemiology, host resistance, resistance gene utilization and resistance mechanisms, pathogen population genetic structure, pathogenicity mechanisms, effector proteome, integrated management, as well as future perspectives, etc. Of course, not all the countries had the same level of work in various areas of research and thus the chapters could not be coherent to this extent to give details about all the topics at the same level.

There are many publications on wheat yellow rust from these countries. We tried to summarize the information based on as many as we could but it was impossible to include all. We express our regret to those whose papers are missed in this book. We especially thank the scientists who made great contributions to the great body of knowledge on yellow rust. We do hope that this book will provide helpful information for interested researchers and prompt future cooperation among these countries to combat wheat yellow rust. Finally, we welcome any comments and criticisms to this book.

Mingju Li
Kunming, China
November 10, 2021

Sajid Ali
Mansehra, Pakistan
November 10, 2021

Contents

Abbreviations used

Abbreviation	Full name
Pst	*Puccinia striiformis* f. sp. *tritici*
APR	Adult-plant resistance
HTSP	High-temperature seedling-plant
HTAP	High-temperature adult-plant
Yr gene	Yellow rust resistance gene
QTL	Quantitative trait loci
GWAS	Qenome-wide association study
DEG	Differentially expressed gene
SPs	Secreted proteins
CYR	Chinese yellow rust
IPM	Integrated pests management
IT	Infection type
FRS	Final rust severity
AUDPC	Area under disease progress curve
ACI	Average coefficient of infection
DNA	Deoxyribonucleic acid
PCR	Polymerase chain reaction
RFLP	Restriction fragment length polymorphism
RAPD	Random amplified polymorphic DNA
SCAR	Sequence characterized amplified region
SSR	Simple sequence repeats
AFLP	Amplified fragment length polymorphism
SNP	Single nucleotide polymorphism

Chapter 1

Wheat Yellow Rust in China–Current Status and Future Perspectives

Mingju Li[1]*, Haoxing Li[2], Chi He[1]

[1]Yunnan Key Laboratory of Green Prevention and Control of Agricultural Transboundary Pests, Agricultural Environment and Resources Institute, Yunnan Academy of Agricultural Sciences, Kunming 650205, China;

[2]State Key Laboratory of Agricultural Microbiology, College of Life Science and Technology, Huazhong Agricultural University, Wuhan 430070, China

* Corresponding author: lily69618@163.com

Abstract: Yellow (stripe) rust, caused by *Puccinia striiformis* f. sp. *tritici*, is considered to be the current most important crop disease in China. Comprehensive research on yellow rust has been carried out among a nationwide network of colleagues working for more than six decades. This review summarized the current knowledge of the epidemiology of wheat yellow rust, including over-summering, over-wintering, inter-regional dispersal of inoculum, resistance gene characterization, utilization and resistance mechanisms, race monitoring and population genetics of the pathogen, pathogenicity mechanisms and effector proteome. The information would be useful to propose a sustainable control strategy in China which mainly base on the use of resistance cultivars, supplemented by ecological regulation and accurate chemicals control.

Keywords: Wheat Yellow (Stripe) Rust; *Puccinia striiformis* f. sp. *tritici*; *Yr* Gene; Physiological Race; China

1.1 Main Agro-ecological Features of China

China located in East Asia, with a land area of about 9.6 million km^2, ranks the third largest country in the world, next only to Russia and Canada. From north to south, the territory of China measures some 5,500 km from latitude 53° 30′ N to 4° N. From west to east, the country extends about 5,200 km from longitude 73° 40′ E to 135° 05′ E, with a time difference of over four hours. China has land borders of 22,800 km long, with 15 contiguous countries: Korea, Mongolia, Russia, Kazakhstan, Kirghizstan, Tajikistan, Afghanistan, Pakistan, India, Nepal, Sikkim, Bhutan, Vietnam, Laos and Myanmar (http://www.china.org.cn/e-china/geography/index.htm).

China's topography is diverse and complicated, with towering mountains, basins of different sizes, undulating plateaus and hills, and flat and fertile plains. The altitude ranges from −154m (Turpan Basin in Xinjiang Province) to 8,848 m (Mt Qomolangma or Mt Everest, the world's highest peak and the main peak of the Himalayas).

China has a marked continental monsoonal climate characterized by great variety. Northerly winds prevail in winter, while southerly winds reign in summer. The four seasons are quite distinct. The rainy season coincides with the hot season. From September to April the following year, the dry and cold winter monsoons from Siberia and Mongolia in the north gradually become weak as they reach the southern part of the country, resulting in cold and dry winters and great differences in temperature. The summer monsoons last from April to September.

The warm and moist summer monsoons from the oceans bring abundant rainfall and high temperatures. China's complex and varied climate results in a great variety of temperature belts, and dry and moist zones. In terms of temperature, the nation from south to north can be divided into equatorial, tropical, sub-tropical, warm-temperate, temperate, and cold-temperate zones; in terms of moisture, from southeast to northwest it can be divided into humid (32% of land area), semi-humid (15%), semi-arid (22%) and arid zones (31%).

The temperature difference between north and south is quite large in winter, the coldest month in the whole year is January, in which the average temperature ranges from −30.9℃ to 22.9℃. The warmest month is July, in which the average temperature ranges from 5.4℃ to 35℃. Annually rain fall ranges from 50 mm to 1600 mm (http://www.china.org.cn/e-china/geography/index.htm).

China has been identified as one of the 17 mega-diverse countries in the world (Mittermeier, 1988), also the most biodiversity-rich country in the Northern Hemisphere, with an ancient creatures origin and plentiful germplasm of cultivated plants and domestic animals. Especially South Central China, including Eastern Himalaya and the Mountain of Southwest China, is one of the 25 biodiversity hotspots globally (Myers et al., 2000).

1.2 The Economic Importance of Wheat Yellow Rust and Historical Epidemic

China is the largest country for wheat production and consumption in the world, annual production accounts for 17% of the world. Wheat is the second most important cereal and food crop in China currently. The total cultivated area is 24.27 million hectares and production is 131.44 million tons (From National Bureau of Statistics, 2018), which plays an important role in both domestic and global food security. While, yellow rust, caused by *Puccinia striiformis* f. sp. *tritici* (*Pst*), is the most destructive disease on wheat, which causes significant yield and economic losses and severely threatens the security of wheat production due to its wide transmission and high frequent epidemics. It has been announced as the most important crop disease nationally by the Ministry of Agriculture and Rural Affairs of China on Sept 15th, 2020.

Historically, the earliest record of wheat yellow rust in China was in an earliest Chinese book, *Ancient Chinese Agricultural Practices*, written by Jia Sixie during Northern Wei Dynasty (533-544), it is about 1,480 years ago. More recently it has been recorded in the book *Agricultural Proverbs* in 1836. In both books, yellow rust was called Jaundice, like a human disease. The former book recorded that the yellow rust was related to spring rainy condition (Li and Zeng, 2002; Chen and Kang, 2017).

Currently, China is regarded as the largest epidemic zone in the world with more than 20 million hectares of wheat crop infected by the disease every year since 1950s (Wan et al., 2007; Chen et al., 2009), the first outbreak of yellow rust disease, which caused yield losses up to 6.0 million tons. Since then, several epidemics occurred in 1964, 1990, 2002, 2017, 2019, and 2020, and annual yield losses averaged about 1 million tons. The disease is more prevalent in the winter-wheat growing areas of the northwest, southwest, and north, and the spring-wheat growing areas in the north-west of China, covering 16 provinces (Li and Zeng, 2002).

Yunnan located in southwest of China, bordered on Myanmar to the west, neighboring the Himalayas, with currently 339,200 hectares wheat cultivated (From National Bureau of Statistics, 2018). Complex terrain and physiognomy, quite different altitudes, as well as diverse climate have resulted in year-round wheat cultivation. This year-around wheat provides green bridge for rust pathogen, which makes the pathogen over-summering and over-wintering easily and infect wheat constantly (Li et al., 2021). *Berberis* spp. and *Mahonia* spp., the alternate hosts of *Pst* (Jin et al., 2010; Zhao et al., 2013; Wang and Chen, 2013), are distributed widely in Yunnan Province (Ying et al., 2011; Fig. 1.1). Collectively these factors have caused frequent epidemic in the past, and Yunnan acts as an epitome of China. The outbreak of wheat yellow rust in Yunnan, not only causes local yield losses, but also provide original incursion sources for other parts of China (Li et al., 2021), which plays a crucial role for large scale disease epidemics.

Comprehensive research on the epidemiology of wheat yellow rust has been carried out among a nationwide network of colleagues working for more than six decades. These studies attempted to understand the over-summering, over-wintering, inter-regional dispersal of inoculum, resistance gene utilization and resistance mechanisms, race monitoring and population genetics, pathogenicity mechanisms, and integrated management in China. The progress of each aspect is introduced as follows.

Fig. 1.1 Wheat cultivation under diverse agro-ecological zones and *Berberis* spp. infected by yellow rust pathogen naturally in China

1.3 Epidemiology

Wheat yellow rust can be dispersed for hundreds or thousands of kilometers by upper air-flow infecting a large area of wheat. Studies indicated that *Pst* over-summers in western wheat growing regions, and over-winters in eastern wheat growing regions in China. The pathogen, thus complete its life-cycle through winter and summer by dispersal between western and eastern China.

As the over-summering is the weakest part in the life-cycle of *Pst*, the over-summering regions are the core areas in epidemic zone, it has been paid much attention and studied in details. Several decades research discovered four over-summering areas in China, namely northwestern (Shaanxi, Gansu, Sichuan, Ningxia and Qinghai), southwestern (Tibet, Yunnan and Guizhou), Xinjiang, and northern (Shanxi and Hebei) over-summering regions (Li and Zeng, 2002; Ma et al., 2005). The northwestern and southwestern regions are the largest and the most important over-summering areas in China, and may overlap. These areas are usually mountainous, wheat grown from lowland (alt. 800 m) to highland (alt. 2,500 m) (Wan et al., 2007). The harvest period extends for 5 months from early May to early October. Sowing dates vary from late September to late October. Autumn-sown lowland wheat emerges before the harvest of highland wheat. Especially for Yunnan province, wheat grown all over the year, which means that host plants are available for the pathogen all year round in these over-summering regions.

Early studies showed that ideal environmental conditions in these northwestern and southwestern areas made them major centers of yellow rust epidemics and the source for new races. Almost all Chinese races were first detected in these regions (Wan et al., 2007), which serve as sources of inoculum and contribute to the development of new races for the other regions of China. The first "breakdown" of every resistance gene has occurred in these regions.

In contrast, the eastern plain regions, including Shaanxi, Shanxi, Henan, Hebei, Hubei, and Shandong provinces, due to high temperature and harvest time, there is no chance for the pathogen to over-summer, and therefore considered as non-over-summering regions.

Wind-borne urediniospores arriving in non-over-summering areas in eastern China can infect autumn-sown wheat seedlings when weather conditions are favourable and susceptible cultivars are grown. Under favourable conditions, the infections continue and produce urediniospores before the pathogen undergoes over-wintering.

During winter, the pathogen survives mainly as mycelium in live plant tissue, but may also survive as urediniospores under certain climatic conditions, e.g., under snow cover or in milder winters (Li and Zeng, 2002). The over-wintered urediniospores and mycelium contribute to spring and summer epidemics and regional spread.

For some mountainous western areas, the pathogen can complete its life cycle locally such as Yunnan, Tibet, West Sichuan, South Gansu, Middle and Southwest Xinjiang, where the *Pst* can oversummer and overwinter (Ma et al., 2005). This increases the chances for mutation and thus the break-down of resistance in newly introduced cultivars, followed by serious epidemics. The achievements of epidemiology have provided scientific basis for effective control strategies for wheat yellow rust.

1.4 Resistance of Chinese Wheat Cultivars

Both major gene-based and partial resistance were used in wheat varieties. Due to the fact that major genes resistance is easily overcome by the pathogen, partial resistance conferring adult-plant resistance (APR), has been paid more attention in wheat production since 1970s. Some APR cultivars such as Libellula, Strampelli, San pastore, Bulgarian 10, Bulgarian 14, Dongfanghong 3, Handan Youzimai, Nongda 168, Nongda 198, Nongda 4356, Nongda 6085, Pingyuan 50, Xuzhou 8, Shaanxi Mazhamai, Qishan Mazhamai, Wangshuibai, Wuming 7, Xiannong 4, Qingnong 1, Qingnong 3, Zhongliang 5, Xiaoyan 6 were widely used and played an important role in Chinese wheat production (He et al., 2011).

During the past decades, 80% Chinese commercial cultivars were susceptible, all-stage resistant cultivars were less than 5% in hotspots areas. During 2003-2008, among 330 commercial cultivars and 194 breeding lines evaluated, only 3.2% cultivars showed all stage resistance, 20.0% showed APR, and 5.7% showed slow-rusting (Zeng et al., 2014). Among 74 Chinese wheat commercial cultivars assessed, only four cultivars showed APR, ten cultivars showed MS-MR at adult stage, the other 60 cultivars showed susceptibility both at seedling and adult stages (Xue et al., 2014).

Trap nursery survey in the year of 2019-2021, recent severe epidemic year in the Chinese yellow rust hotspot Yunnan, indicated that most of Chinese commercial wheat cultivars showed highly susceptible in the field, only a few of them showed good APR, such as, Zhoumai 22, Lantian 31, etc. and Yunnan local released cultivars such as Yunmai 34, Yunmai 56, Yunmai 68, Yunmai 101, Yunmai 110, Yunmai 112, etc. Further efforts will be required for developing new resistant cultivars with diverse genetic background.

1.5 Major Resistance Gene Exploitation and their Effectiveness

Resistance genes to *Pst* can be broadly categorized into two main classes, namely, major and minor resistance genes. The major gene-based resistance can be called all-stage resistance. Minor resistance genes also called partial, non-race specific, slow-rusting, durable or adult plant resistances (Rosewarne et al., 2013).

Till now, 80 *Yr* genes (denoted from *Yr1* to *Yr80)* have been formally designated and approximately 100 *Yr* genes temporarily designated (McIntosh et al., 2017; Chen and Kang, 2017). Among these, *Yr11, Yr12, Yr13, Yr14, Yr16, Yr18, Yr29, Yr30, Yr34, Yr46, Yr48, Yr49, Yr58, Yr60, Yr68, Yr71,* and *Yr75* confer APR, and *Yr36, Yr39, Yr52*, *Yr59*, *Yr62* confer higher-temperature adult-plant (HTAP) and *Yr54*, *Yr77, Yr78, Yr79, Yr80* are quantitative trait loci (QTL) conferring resistance (Han et al., 2018). The others confer all-stage resistance.

Both gene postulation and molecular markers are used to characterize *Yr* gene in China. Currently, some molecular markers for important genes such as *Yr1*, *Yr2, Yr5, Yr9, Yr10, Yr15, Yr17, Yr18, Yr24,* and *Yr26* have been developed and used to test these *Yr* genes (Table 1.1). The others genes were postulated by a series of *Pst* strains and known *Yr* gene cultivars.

Table 1.1 *Yr* genes linked markers, their primer sequences and references

Yr gene	Linked markers (cM)	Sequence (5′-3′)	Linked gene (cM)	References
Yr1	SSR Marker Xgwm372	AATAGAGCCCTGGGACTGGG GAAGGACGACATTCCACCTG	*Sr48* (16.5)	Xu et al., 2016
	SSR Marker Xgwm382	GTCAGATAACGCCGTCCAAT /CTACGTGCACCACCATTTTG		Bansal et al., 2009; Xu et al., 2016

(continuous)

Yr gene	Linked markers (cM)	Sequence (5′-3′)	Linked gene (cM)	References
Yr2	SSR Marker wmc364	ATCACAATGCTGGCCCTAAAAC CAGTGCCAAAATGTCGAAAGTC	—	Xu et al., 2016
Yr5	EST-STS Marker STS9/STS10	AAAGAATACTTTAATGAA /CAAACTTATCAGGATTAC	*YrSP* (12.6), *Yr43* (65.5), *Yr44* (41.5), *Yr53* (35.6)	Chen et al., 2003; Murphy et al., 2009; Zhang et al., 2009
	SCAR Marker S1320	CAATAGTTAGGCAAATTACATCG /TGCAAAGTACCTCATTTTGAGAA		Dong et al., 2019
	S19M93	TAATTGGGACCGAGAGACG/ TTCTTGCAGCTCCAAAACCT		Smith et al., 2007
	Yr5-Insertion	CTCACGCATTTGACCATATACAAcT /TATTGCATAACATGGCCTCCAGT		Marchal et al., 2018
Yr9	H20	GTTGGAAGGGAGCTCGAGCTG /GTTGGGCAGAAAGGTCGACATC	*Sr31, Lr26*	Cheng et al., 2008; McIntosh et al., 1995
	P6M12-P	GTACTAGTATCCAGAGGTCACAAG /CAGACAAACAGAGTACGGGC		Mago et al., 2005
	SCAR Marker AF1/AF4	GGAGACATCATGAAACATTTG /CTGTTGTTGGGCAGAAAG		Francis et al., 1995
Yr10	SSR Marker Xpsp3000 (1.2)	GCAGACCTGTGTCATTGGTC /GATATAGTGGCAGCAGGATACG	Brown glune (2.03)	Wang et al., 2002
	SCAR Marker SC200	CTGCAGAGTGACATCATACA /TCGAACTAGTAGATGCTGGC		Shao et al., 2001
	Yr10- RT	5'-CACTTGAGGTATCTGAGTCTAG /5'-CAATGAACACCAGTTGTCCTA		Liu et al., 2014
	SCAR Marker E51100	TCAAGGAGGTCAGTGACAG /TCAGGGAGGTGTAGCCTAAT		Liu et al., 2014
Yr15	SSR Marker Barc8 (0.0)	GCGGGAATCATGCATAGGAAAACAGAA /GCGGGGGCGAAACATACACATAAAAACA	*Yr24, Yr64* (21.7), *Yr65* (14.1)	Murphy et al.,2009; Lesleyr et al., 2009
	SSR Marker Xgwm273 (0.4)	ATTGGACGGACAGATGCTTT /AGCAGTGAGGAAGGGGATC		Peng et al., 2000; Cheng et al., 2014
	Y15K1-F2/ uhw301R	GGAGATAGAGCACATTACAGAC /TTTCGCATCCCACCCTACTG		Klymiuk et al., 2018
Yr17	SC-385	CTGAATACAAACAGCAAACCAG /ACAGAAAGTGATCATTTCCATC	*Lr37, Sr38*	Helguera et al., 2003
	VENTRIUP/LN2	AGGGGCTACTGAACCAAGGCT /TGCAGC TACAGC AGTATGTACACAAAA		Helguera et al., 2003
Yr18	EST-STS Marker cs-LV34	GTTGGTTAAGACTGGTGATGG /TGCTTGCTATTGCTGAATAGT	*Lr34, Sr57, Pm38, Ltn*	Lagudah et al., 2006
	L34 DINT9F/ L34PLUSR	TTGATGAAACCAGTTTTTTTTCTA /GCCATTTAACAtAATCATGATGGA		Lagudah et al., 2009
Yr24	SSR Marker Xgwm273	ATTGGACGGACAGATGCTTT / AGCAGTGAGGAAGGGGATC	*Yr15, Yr64, Yr65*	Liu et al., 2005; Cheng et al., 2014

(continuous)

Yr gene	Linked markers (cM)	Sequence (5′-3′)	Linked gene (cM)	References
Yr26	SSR Marker Barc181 (6.7)	CGCTGGAGGGGGTAAGTCATCAC /CGCAAATCAAGAACACGGGAGAAAGAA	*Yr24*	Wang et al., 2008
	EST-STS Marker We173 (1.4)	GGGACAAGGGGAGTTGAAGC /GAGAGTTCCAAGCAGAACAC		Wang et al., 2008
	EST-STS Marker CON-4	GTGCTGTACCTGACGACGGA /GTGGAGATGTTGGGCTTGG		Zhang et al., 2013
	Xgwm11 (1.9)	GGA TAGTCAGACAAT TCT TG TG GTGAA TTG TG TCT TG TA TGC TTCC		Roder et al., 1998; Liu et al., 2008
	Xgwm18	TGCGCCA TGAT TGCA TTA TCT TC GGT TGC TGA AGAACC TTA TT TAGG		Roder et al., 1998; Liu et al., 2008

Table 1.2 Resistance genes identified in various wheat varieties/lines using molecular markers and/or gene postulation techniques

Reference	Lines tested	Lines	Method used	*Yr* genes tested	*Yr* genes reported
Li et al., 2008	Commercial varieties, breeding lines	126	Molecular markers	*Yr9*, *Yr5*, *Yr10*, *Yr15*, *Yr26*	*Yr9* (42%), *Yr5* (5%), *Yr10* (3%), *Yr15* (3%), *Yr26* (2%)
Liu et al., 2008	Commercial varieties, breeding lines	239	Molecular markers	*Yr26*	*Yr26* (15%)
Yang et al., 2008	Commercial varieties	231	Molecular markers	*Yr18*	*Yr18* (6%)
	Landraces	422			*Yr18* (85%)
Wang et al., 2011	Commercial varieties	137	Molecular markers	*Yr9*, *Yr10*, *Yr15*	*Yr9* (16%), *Yr10* (3%), *Yr15* (14%)
Li et al., 2011	Commercial varieties	52	Gene postulation	*Yr9*, *Yr21*, *Yr8*, *Yr6*, *Yr17*, *Yr26*, *Yr7*, *Yr27*	*Yr9* (25%), *Yr21* (8%), *Yr8* (6%), *Yr6* (4%), *Yr17* (4%), *Yr26* (4%), *Yr7* (2%), *Yr27* (2%)
Lv et al., 2013	New wheat lines	30	Gene postulation	*Yr21*, *Yr27*, *Yr1*	*Yr21* (33%), *Yr27* (20%), *Yr1* (17%)
Zhang et al., 2014 a	Commercial varieties	75	Gene postulation	*Yr3*, *Yr2*, *Yr9*, *Yr1*, *YrSp*	*Yr3* (27%), *Yr2* (20%), *Yr9* (16%), *Yr1* (14%), *YrSp* (14%)
Zhang et al., 2014 b	Commercial varieties	75	Molecular markers	*Yr10*, *Yr18*, *Yr9*	*Yr10* (17%), *Yr18* (1%), *Yr9* (33%)
Xu et al., 2018	Commercial varieties	22	Molecular markers	*Yr1*, *Yr10*, *Yr24*	*Yr1* (50%), *Yr10* (9%), *Yr24* (0)
Dong et al., 2019	Germplasms	348	Molecular markers	*Yr5*, *Yr10*, *Yr18*	*Yr5* (35%), *Yr10* (23%), *Yr18* (2%)
Xu et al., 2019	Commercial varieties	70	Gene postulation	*Yr21*, *Yr3*, *Yr1*, *Yr5*, *YrSD*, *Yr32*, *Yr10*, *YrSu*, *Yr26*	*Yr21* (26%), *Yr3* (29%), *Yr1* (16%), *Yr5* (11%), *YrSD* (4%), *Yr32* (4%), *Yr10* (9%), *YrSu* (1%), *Yr26* (1%)
			Molecular markers	*Yr5*, *Yr9*, *Yr10*, *Yr15*, *Yr26*	*Yr5* (16%), *Yr9* (3%), *Yr10* (13%), *Yr15* (1%), *Yr26* (6%)

Research indicated that just fewer genes such as *Yr1*, *Yr2*, *Yr3, Yr9*, have been widely used in Chinese wheat varieties. In recent years, cultivars carrying *Yr10*, *Yr26* have been released more than before. By now the resistance of these genes have been brokendown. The other genes have only a low frequency in Chinese commercial wheat cultivars. Table 1.2 described the probable *Yr* genes frequency in Chinese commercial wheat cultivars, wheat landraces and breeding lines. Virulence monitoring and field trap nursery showed that only *Yr5, Yr15* and *Yr18* expressed good resistance in China at present. Just a small number of Chinese commercial cultivars carry these genes, and needs to be incorporated into the future Chinese lines.

Yr18 (*Lr34*/*Yr18*/*Sr57*/*Pm38*/*Ltn1*) is a very important slow-rusting resistance gene which conferred pleiotropic APR, to yellow rust, leaf rust, stem rust, powdery mildew, and conferring the phenotypic marker of leaf tip necrosis (Rosewarne et al., 2013), and has been widely used in many countries (Ali et al., 2017, 2018). Molecular characterization of 231 Chinese wheat commercial cultivars and 422 landraces indicated that only 6.1% commercial cultivars and 85.1% landraces carried *Yr18*. The Southwest Winter Wheat Region has a higher frequency than that of the whole China (Yang et al., 2008). Thus, developing slow-rusting cultivars from landraces will be a promising aspect in future.

1.6 Quantitative Trait Loci (QTL) and Genome-wide Association Study (GWAS) of Yellow Rust Resistance Genes

More than 140 QTLs for yellow rust resistance in wheat and 47 chromosomal regions were identified worldwide (Rosewarne et al., 2013). Many of the regions contain more than one gene, for example, chromosome 2B has a region that contains a number of major genes such as *Yr27*, *Yr31* and *Yr7*, as well as a number of QTLs that are effective at adult plant stage. The strongest effect QTLs are generally associated with major genes such as *Yr17*. Many of these QTLs are identical genes. Lots of work have been carried out in China to identify QTL of yellow rust resistance, using Chinese wheat landraces and commercial cultivars which exhibited APR to yellow rust over many years.

Several Chinese wheat landraces such as Humai 15, ‘Guangtoumai’, Pingyuan 50, and Qing Shumai, have been studied for their QTLs. In Humai 15, a major effect QTL was identified and was located on the centromere of chromosome 2B, accounted for 47.2% of the phenotypic variation (Yuan et al., 2018). In Pingyuan 50, three QTLs on chromosomes 2BS, 5AL, and 6BS, explaining 4.5%~19.9% of the phenotypic variation respectively, were designated as *QYr.caas-2BS*, *QYr.caas-5AL*, and *QYr.caas-6BS* (Lan et al., 2010). In ‘Guangtoumai’, a major locus named *QYr.GTM-5DL* located on chromosome 5DL, explained up to 44.4% of the total phenotypic variation, and was present in 5.3% of the 247 surveyed Chinese wheat landraces (Wu et al., 2021). In Qing Shumai, a major QTL on chromosome 6D was designated as *QYr.cau-6DL*, with which a SSR marker (gpw5179) was tightly linked (Zhang et al., 2017).

Commercial wheat cultivars such as Bainong 64, Chuanmai 42, Chuanmai 55, Xinong1376, Qinnong 142, and Shannong 33 also have been studied. Five loci were detected from Bainong 64, and were located at the same positions as the *Yr29/Lr46* and *Yr46/Lr67* genes, respectively (Ren et al., 2012). Locus *Qyr.saas-7B* from Chuanmai 42, and *Qyr.saas-1B* and *Qyr.saas-2A* from Chuanmai 55 was identified. The *Qyr.saas-1B* explained 6.24%~34.22% of the phenotypic variation, overlapped with *Yr29*, whereas a significant additive effect was observed when all three QTLs were combined (Yang et al., 2019). Six QTL from Xinong 1376 identified, *QYr.nwafu-4AL* and *QYr.nwafu-6BL.3* conferred stable resistance in all environments (Mu et al., 2019a). Four QTLs in Qinnong 142 on chromosome arms 1BL, 2AL, 2BL, and 6BS were characterized (Zeng et al., 2019). Four consistent QTL were detected on 1BL, 2AS, 3DL, and 6BS in Shannong 33. The 2AS locus was identified as *Yr17*. The QTL identified on 1BL and 6BS likely correspond to *Yr29* and *Yr78*, respectively. A QTL on 3DL explaining 5.8%~12.2% of the phenotypic variation is likely to be new (Huang et al., 2020). Several other studies have been done to identify QTLs in diverse wheat (Lu et al., 2009; Wu et al., 2017; Wu et al., 2018; Yao et al., 2019; Mu et al., 2019b; Ye et al., 2019; Cheng et al., 2020).

All these QTL loci and their closely linked molecular markers will facilitate marker-assisted selection, gene pyramiding, and improve the level of APR against yellow rust in breeding programs.

1.7 Understanding the Mechanism of Resistance

Several studies have been undertaken in China to understand the mechanism of resistance. Gene *TaRLP1.1* was characterized which greatly contributed to the hypersensitive response during the pathogen-host interaction (Jiang et al., 2003). A calcium binding EF-hand protein 1 gene (*TaCab1*) was involved in the plant-pathogen recognition and symptom development (Feng et al., 2011). Gene *TaMYB4* significantly upregulated at the early stage and 48 h after inoculation with the incompatible *Pst* (Al-Attala et al., 2014). The transcript of gene *TaDIR1-2* significantly induced during the compatible interaction of wheat with *Pst* (Ahmed et al., 2017). Tao et al. (2018) found that the majority of differentially expressed genes (DEGs) were up-regulated in high-temperature seedling-plant (HTSP) resistance. Zhang et al. (2018) demonstrated that wheat Nuclear Transport Factor 2 (*TaNTF2*) was a contributor for wheat resistance to *Pst*. QRT-PCR verified the up-regulated expression of *TaNTF2* in response to avirulent *Pst*. Mamun et al. (2018)

reported an Autophagy-related 8 gene *TaATG8j*, which contributed to resistance by regulating cell death. Chen et al. (2019) deduced that wheat with yellow rust resistance could maintain high resistance and photosynthetic capacity by regulating the antioxidant system, disease-resistance related enzymes and genes, and the levels of photosystem II reaction center proteins. *TaSTP6* is an active sugar transporter in wheat, which exhibit enhanced expression in leaves upon infection by *Pst*. Huai et al. (2019) found that *Pst* infection stimulated biosynthesis of accumulation of abscisic acid (ABA) in host cells and thereby upregulated *TaSTP6* expression, which increased sugar supply and promotes fungal infection. Zhang et al. (2019) reported that *YrAS2388* only existed in *Aegilops tauschii*, mutation of the *YrAS2388R* allele disrupted its resistance to *Pst* in synthetic hexaploid wheat, transgenic plants with *YrAS2388R* show resistance to eleven *Pst* races in common wheat. Yang et al. (2020) identified *TaSBT1.7* induced in wheat leaves by chitin and flg22 elicitors, as well as six examined pathogens, implying a role for *TaSBT1.7* in plant defense. Wang et al. (2019) cloned a leucine rich repeat (LRR)-RLK gene, *TaXa21*, and found that *TaXa21* functions as a positive regulator of wheat HTSP resistance to *Pst*. The *TaRPM1* transcription was rapidly upregulated upon *Pst* inoculation under high temperature. *Yr36* encodes Wheat Kinase START1 (*WKS1*), an effective HTAP resistance gene and confers resistance to a broad spectrum of *Pst* races (Wang et al., 2020). *WKS1*-mediated *Pst* resistance is accompanied by leaf chlorosis in *Pst*-infected regions. The *WKS1* reduced the rate of photosynthesis, regulate leaf chlorosis, delay the growth of *Pst* pathogen, and confer *Pst* resistance (Wang et al., 2019). The expression level of *TaMCA4* in wheat leaves was sharply upregulated when challenged with the avirulent race of *Pst*, whereas the transcript level was not significantly induced by the virulent race (Wang et al., 2012).

More research on wheat resistance mechanisms to *Pst* pathogen is ongoing and a thorough understanding of this mechanism will facilitate breeding for yellow rust resistance in wheat using biotechnology.

1.8 Pathogen Population Biology and Virulence Variation

As the rapid variation of *Pst* population virulence leads to resistance breakdown of the existing resistant wheat cultivars leading to disease epidemics, study on virulence variation, population genetics and pathogenicity mechanism of pathogen has been paid much attention since 1950s.

The first report about *Pst* race identification was made by Fang (1944), who used seven wheat cultivars and 1 barley cultivars as differentials, and identified 9 races from Yunnan collections. Due to the variation of races, Chinese differentials has been adjusted several times and currently includes 19 cultivars (Table 1.3), and use chronological nomenclature to name races (Hu et al., 2014). Since 1957, the major virulence patterns were named as races of CYR (Chinese yellow rust) and minor virulence patterns were considered as 'pathotypes' with low frequencies and narrow virulence spectrum. Annual virulence variation and frequency of races occurrence were monitored by the National Initiative of Wheat Rust and the results have provided valuable information for breeding programs. Till 2015, 303 races or pathotypes had been identified in the whole China (National Initiative of Wheat Rust Annual Report, 2015).

Most of races in China can be monitored in these hotspots such as South Gansu, Northwest Sichuan and Yunnan, and some new races discovered in these regions firstly, where pathogen can mutate easily and has a high genetic diversity.

The predominant races in each period and its infected genes are listed in Table 1.4. Due to the variation of *Pst* variation, eight large scale cultivars resistance broke down and then replaced by new resistant cultivars since 1950s, to ensure security of wheat production. Each period the responsible races and related cultivars listed in the Table 1.5.

Table 1.3 Source and *Yr* genes present In Chinese differentials used for wheat yellow rust

No.	Cultivars	*Yr*-genes	Source
1	Trigo Eureka	*Yr6*	+
2	Fulhard	+	+
3	Lutesens 128	+	Bulgaria
4	Mentana	+	Italy
5	Virgilio	*YrVir1, YrVir2*	Italy
6	Abbondanza	+	Italy
7	Early Premium	+	America
8	Funo	*YrA*, +	Italy
9	Danish 1	*Yr3*	Denmark
10	Jubilejina II	*YrJu1, YrJu2, YrJu3, YrJu4*	Bulgaria
11	Fengchan 3	*Yr1*	China
12	Lovrin 13	*Yr9*, +	Romania
13	Kangyin 655	*Yr1, YrKy1, YrKy2*	+
14	Suwon 11	*YrSu*	Korea
15	Zhong 4	+	China
16	Lovrin 10	*Yr9*	Romania
17	Hybrid46	*Yr4b, YrH46*	+
18	*T. spelta album*	*Yr5*	+
19	Guinong 22	*YrGui1, YrGui2, YrGui3*	China
20	Mingxian 169	None	China

Note: "+" indicates unknown. Resistance genes and source cited from Chen et al. (2009).

Table 1.4 Predominant race, infected genes, and its highest frequency

Race	Infected genes	Year of named	Highest frequency (%) (year)
CYR1	—	1955	76.9 (1959)
CYR8	—	1958	32.9 (1960)
CYR10	—	1960	51.5 (1964)
CYR13	—	1962	16.3 (1966)
CYR17	*Yr1, Yr6, Yr7, Yr8, YrA, YrSu*	1966	77.6 (1971)
CYR18	*Yr3*	—	38.6 (1972)
CYR19	*Yr1, Yr2, Yr3a, Yr4a, Yr7, Yr8, YrA, YrSu*	1972	88.6 (1979)
CYR21	*Yr1, Yr2, Yr3a, Yr4a ,Yr6, Yr7, Yr8, YrA, YrSu*	1975	—
CYR22	*Yr1, Yr2, Yr3a, Yr4a, Yr6, Yr8, YrA, YrSu*	1975	15.7 (1983)
CYR23	*Yr1, Yr2, Yr3a, Yr4a, Yr6, Yr8, YrA, YrSu, YrGaby, YrRes, YrSel*	1976	27.6 (1980)
CYR25	*Yr1, Yr2, Yr3, Yr8, Yr17, Yr19, YrA, YrGaby*	1978	44.2 (1982)
CYR26	*Yr1, Yr2, Yr3a, Yr4a, Yr7, YrSu, YrA*	1980	24.6 (1984)
CYR27	*Yr1, Yr2, Yr3a, Yr4a, Yr6, Yr8, YrA, YrSu, YrSP, YrAlba*	1983	12.5 (1983)
CYR28	*Yr1, Yr2, Yr3, Yr7, Yr8, Yr9, Yr19, Yr27, YrA, YrAlba, YrCle, YrRes*	1985	7.3 (1986)
CYR29	*Yr1, Yr2, Yr3, Yr8, Yr9, Yr19, Yr27, YrA, YrCle, YrCV, YrGaby, YrSD*	1985	40.3 (1989)
CYR30	*Yr1, Yr2, Yr3, Yr4, Yr8, Yr9, Yr17, Yr19, Yr27, YrA, YrAlba, YrCle, YrCV, YrGaby, YrRes, YrSD*	1991	7.9 (1995)
CYR31	*Yr1, Yr2, Yr3, Yr6, Yr7, Yr9, Yr27, YrA, YrAlba, YrCle, YrCV, YrGaby, YrRes, YrSD, YrSu*	1993	16.7 (1997)
CYR32	*Yr1, Yr2, Yr3, Yr4, Yr6, Yr7, Yr8, Yr9, Yr17, Yr18, Yr27, Yr29, Yr32, YrA, YrAlba, YrCle, YrCV, YrGaby, YrRes, YrSD, YrSP, YrSu, YrSK*	1994	34.6 (2002)
CYR33	*Yr1, Yr2, Yr3, Yr4, Yr6, Yr7, Yr8, Yr9, Yr17, Yr18, Yr27, Yr29, Yr32, YrA, YrAlba, YrCle, YrCV, YrGaby, YrRes, YrSD, YrSP, YrSK*	1997	26.7 (2007)
CYR34	*Yr1, Yr2, Yr3, Yr4, Yr6, Yr7, Yr8, Yr9, Yr10, Yr17, Yr18, Yr24, Yr26, Yr27, Yr29, Yr32, YrA, YrSP, YrSK*	2016	34.2 (2017)

Note: Data compiled based on the results of Hu et al. (2014), Liu et al. (2017). "—" means no data available.

Table 1.5 Replacement of eight large scale cultivas in China since 1950s*

Period	Main cultivars (*Yr* genes) of breakdown resistance	Corresponding races
1950–1960	Bima 1 (*Yr1*), Xibei 54 (*Yr1*), Xibei Fengshou, Nongda 183, etc.	CYR1
1960–1962	Quality, CII 2203, Xibei 134, Shaannong 9, Yupi, Gansu 96, etc.	CYR8, CYR10
1962–1970	Mentana and its derivates.	CYR13, CYR16
1970–1976	Beijing 8 (*Yr1*), Beijing 10, Shijiazhuang 54, Abbondanza, etc.	CYR17, CYR18

(continuous)

Period	Main cultivars (*Yr* genes) of breakdown resistance	Corresponding races
1976–1985	Fengchan 3(*Yr1*), Guancun 1, Taishan 1, Taishan 4, Weidong 8, Nongda 139, Funo (*YrA*), Baiquan 40, Yanda 24, Jubilejina II(*YrJu1, YrJu2, YrJu3, YrJu4*), TianXuan 15, Zhongliang 5, Wunong 132, Qingchun 2, etc.	CYR19, CYR23, CYR25
1985–1995	Lovrin and its derivates: Lovrin 10 (*Yr9*),Lovrin 13 (*Yr9*,+), Fengkang 8 (*Yr9*), Lumai 1 (*Yr9*,+), Lumai 5, 7, 11, Lumai 14 (*Yr9*,+), Lumai 15, Een 1 (*Yr9*), Bainong 3217 (*Yr1,2,3*,+), Xiaoyan 6, Jinmai 33 (*Yr1*,+), Shaan 7859, Wan 7107, Jimai 26, Jimai 30, Jimai 36, Yumai 13, Yumai 18 (*Yr1*,+), Yumai 21(*Yr9*,+), Yumai 25, Yangmai 5, Yangmai 158, etc.	CYR28, CYR29, Lv10, and Lv13 pathotype group
1995–2000	Fan 6 and its derivates, Suwon 11(*YrSu*) lines, Mianyang 11 lines (Mianyang 11, Mianyang 15, Miangyang 19, Mianyang 20), Chuanmai lines [Chuanmai 22 ,Chuanmai 23 (*Yr9*)], etc.	CYR30, CYR31, CYR32, Hy, and Su pathotype group
2001–date	Guinong 21, Guinong 22(*YrGui1, YrGui2, YrGui3*), Lantian 17 (*Yr26*, *Yr26*=*Yr24*), Chuanmai 42 (*Yr24*), 92R137 (*Yr26*) and its derivates, Moro (*Yr10*, *YrMor*), Tao lines: Tao 157, Tao 153, and Tao 718, Qingnong lines: Qingnong 3; Zhongliang 4, Lantian lines: Lantian 3, Lantian 4, and Lantian 6, Zhongliang lines: Zhongliang 16, Zhongliang 17, Zhongliang 18, Zhongliang 20, Zhongliang 25, and Zhongliang 29, Tianxuan lines: Tianxuan 40, Tianxuan 41, and Tianxuan 42, Mianyanglines: Mianyang 26, Mianyang 27, Mianyang 28, and Mianyang 30, Longjian lines: Longjian 64, and Longjian 196, Jimai 30, Beijing 837, Jinmai 47, Lumai 22, Gaoyou 50, Wudu 12, Wudu 15, Ji'nan 17, Wanmai 19, Yumai 34, Yumai 47, Yumai 49, Yumai 54, Yumai 70, Zhengmai 9023, Yanyou 361, Longyuan 061, Longyuan 931, Hybrid 46, Kangying 655, Jimai 20, Zhengmai 366, Shiluan 02-1, Gaocheng 8901, Zhongyou 9507, etc.	CYR32, CYR33, CYR34, Hy, Su, V26, ZS, *Yr10* and G22 pathotype group

Note: Data compiled based on the results of Wan et al. (2004), Wan et al. (2007), Kang et al. (2015) and He et al. (2018).

Recently, Li et al. (2018) used newly established 18 *Yr* single-gene differentials, namely *Yr1, Yr5, Yr6, Yr7, Yr8, Yr9, Yr10, Yr15, Yr17, Yr24, Yr27, Yr32, Yr43, Yr44, YrSP, YrTr1, YrExp2*, and *YrTye (Yr76)*, and octal code nomenclature to identify the *Pst* races from Yunnan collections in 2016. They characterized 64 races from 136 isolates, among them, 12 races appeared over twice, each frequency more than 1%, the samples of the 12 races accounted for 61.76% of total samples identified (Table 1.6), while the others appeared only once. They found that the *Pst* population in Yunnan was highly variable in terms of races and virulence. The top two most frequent races were 550273 (28.68%) and 550073 (11.76%), and the remaining races had frequencies less than 5%. No virulence were found for *Yr5*, *Yr10*, *Yr15*, and *Yr32*. The virulence frequencies to *Yr24*, *YrTr1*, *Yr8*, and *Yr17* ranged from 0.74% to 11.76%. The virulence frequency to *Yr27* was 52.94%; and those to *Yr1*, *Yr6*, *Yr7*, *Yr9*, *Yr43*, *Yr44*, *YrSP*, *YrExp2*, and *YrTye (Yr76)* ranged from 79.94% to 91.91%. The virulence frequency of *Pst* population to*Yr5*, *Yr10*, *Yr15*, *Yr24*, *Yr32,* and *YrTr1* between 2016 and 2018 was low, suggesting these genes as effective. The remaining genes were considered as non-effective currently in Yunnan Province (Fig. 1.2).

Table 1.6 Races identified and its virulence, frequency and distribution in Yunnan in 2016

No.	Race	Infected *Yr*-gene	Frequency (%)
1	550273	*Yr1, Yr6, Yr7, Yr9, Yr27, Yr43,Yr 44, YrSP, YrExp2, YrTye*	28.68
2	550073	*Yr1, Yr6, Yr7, Yr9, Yr43, Yr44, YrSP, YrExp2, YrTye*	11.76
3	540273	*Yr1, Yr6, Yr7, Yr27, Yr43, Yr44, YrSP, YrExp2, YrTye*	4.41
4	550272	*Yr1, Yr6, Yr7, Yr9, Yr27, Yr43, Yr44, YrSP, YrExp2*	3.68
5	550233	*Yr1, Yr6, Yr7, Yr9, Yr 27, Yr44, YrSP, YrExp2, YrTye*	2.21
6	570073	*Yr1, Yr6, Yr7, Yr8, Yr9, Yr43, Yr44, YrSP, YrExp2, YrTye*	2.21
7	550003	*Yr1, Yr6, Yr7, Yr9, YrExp2, YrTye*	1.47
8	550031	*Yr1, Yr6, Yr7, Yr9, Yr44, YrSP, YrTye*	1.47
9	550033	*Yr1, Yr6, Yr7, Yr9, Yr44, YrSP, YrTye*	1.47
10	550072	*Yr1, Yr6, Yr7, Yr9, Yr43, Yr44, YrSP, YrExp2*	1.47
11	550271	*Yr1, Yr6, Yr7, Yr9, Yr43, Yr44, YrSP, YrTye*	1.47
12	561262	*Yr1, Yr6, Yr7, Yr8, Yr17, Yr27, Yr43,Yr 44, YrExp2*	1.47
Total			61.76

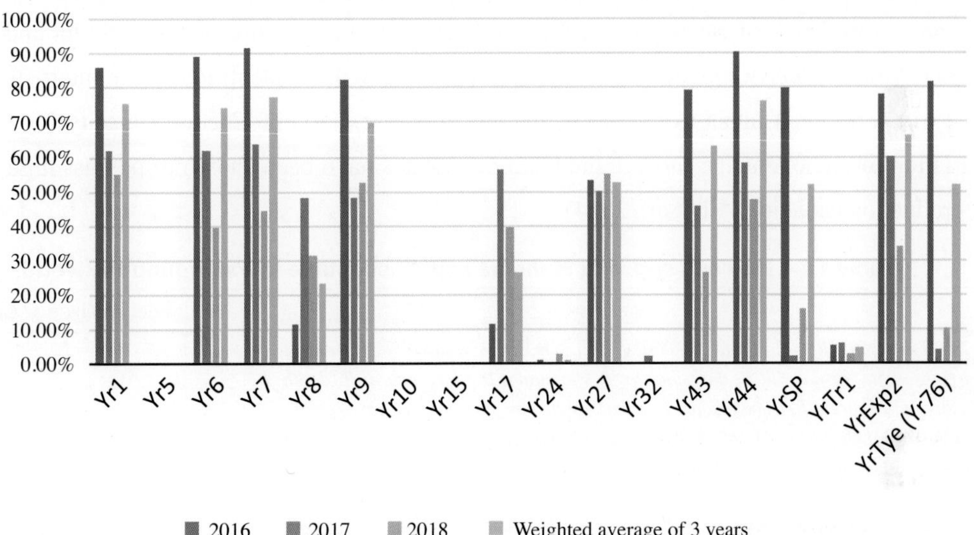

Fig. 1.2 Virulence frequency of *Pst* population to 18 *Yr* genes in Yunnan between 2016 and 2018

The variation in *Pst* originates through mutation, heterokaryosis and sexual reproduction. The mutation is an important way for *Pst* variation in China (Shang et al., 1994). *Pst* may produce new strains by heteronuclear interaction and new strains of multinuclear types (Kang et al., 2015).

Ali et al. (2014) has identified the extended Himalayan regions as the center of diversity for *Pst*, while Jin et al. (2010) proved that *Berberis* serve as an alternative host for *Pst*. Several studies were oriented towards understanding the role of sexual reproduction on pathogen population structure. Wang and Chen (2015) found that Oregon grape (*Mahonia aquifolium*) can serve as an alternate host for the *Pst* under artificial inoculation. Zhao et al. (2013) isolated 20 *Pst* strains from different naturally infected *Berberis* species in Gansu, Shaanxi, Tibet, and got the evidence of sexual reproduction for the first time and proved sexual reproduction caused variation of pathogenicity. Till now, 37 *Berberis* spp. and one *Mahonia* species has been identified as alternative host for *Pst* in China under controlled or natural conditions (Mehmood et al., 2020; Rodriguez-Algaba et al., 2021; Li et al., 2021), the details are given in Table 1.7. It has been proved that *Pst* could not only infect *Berberis* spp. naturally to fulfil its sexual reproduction, but also play an important role in yellow rust disease cycle, which provides inoculum for disease epidemic and may cause new variation of pathogenicity. Sexual reproduction could be the explanation to the high diversity in the natural *Pst* population.

There are estimated 500 *Berberis* spp. and 60 *Mahonia* spp. documented around the world, approximately half of them are native to China, while near 100 *Berberis* spp. and 16 *Mahonia* spp. are grown in Yunnan (Ying et al., 2011). Recent research based on weather data found that China has mid-risk of *Berberis* infection by *Pst*, with a risk score of 0.67 (Sinha and Chen, 2021). More *Berberis* spp. and *Mahonia* spp. probably will be identified as the alternative host of *Pst*. Therefore, Yunnan as well as China has higher sexual reproduction capacity for *Pst*, and variation of pathogenicity. Indeed, the Chinese isolates have been shown to possess high sexual reproduction capacity (Ali et al., 2014).

Table 1.7 *Berberis* spp. and *Mahonia* spp. susceptible to *Pst* around the world

Species	No.	Infection type	Country
B. aggregata, B. aggregata var. integrifolia, B. approximate, B. atrocarpa, B. circumserrata, B. davidii, B. dasystachya, B. diaphana, B. dictyoneura, B. dubia, B. ferdinandi-coburgii, B. franchetiana, B. franchetiana var. glabripes, B. guizhouensis, B. gyalaica, B. heteropoda, B. jamesiana, B. jaeschkeana var. bimbilaica, B. julianae, B. kansuensis, B. kaschgarica, B. kongboensis, B. lepidifolia, B. nummularia, B. phanera, B. platyphylla, B. poiretii, B. pruinosa, B. stenostachya, B. vernae, B. wangii, B. wilsonae, B. zanlanscianensis	33	Artificial	China
B. brachypoda, B. potaninii, B. shensiana, B. soulieana	4	Artificial + Natural	China
Mahonia aquifolium	1	Artificial	China
B. lycium, B. orthobotrys, B. pseudumbellata, B. stewartiana, B. brandisiana, B. pseudumbellata subsp. pseudumbellata, B. pseudumbellata subsp. gilgitica	7	Artificial	Pakistan
B.chinensis, B. holstii, B. koreana, B. vulgaris, B. thunbergii	5	Artificial	US
B. vulgaris subsp. Seroi, B. vulgaris subsp. australis	2	Artificial	Spain

Note: Cited from Mehmood et al. (2020), Rodriguez-Algaba et al. (2021) and Li et al. (2021).

1.9 Molecular Population Genetic Structure

Molecular genotyping and subsequent population genetic analysis have been done on Chinese populations with an aim to understand the diversity and migration across various epidemic zones, and revealed that *Pst* population in China existed high genetic diversity, especially for the over-summering regions, such as South Gansu, Northwest Sichuan, and Yunnan (Shan et al., 1998; Wang et al., 2010; Chen et al., 2013; Kang et al., 2015; Alam et al., 2021; Li et al., 2021). Various set of markers has been used for this purpose, including RAPD, SSRs, PSR, EST-SSRs and AFLP (Chen, 2008; Lu et al., 2009; Duan et al., 2010).

Table 1.8 Primers developed to detect *Pst* SNP loci and its information

Gene	Protein name	Species	GeneBank accession no.	Primers (5'-3')	Product length (bp)	No. of SNP loci
Cdc2	Cyclin dependent kinase 2	*Pst*	GQ911579.1	Cd28S:AAATCATCCACATCTGCTCCAC Cd352A:AAATCATCCACATCTGCTCCAC	325	3
Ef-1α	Elongation factor 1 α	*Pgt*	X73529.1	Ef137S:AAGCCGCATCCTTCGTTG Ef531A:TTGCCATCCGTCTTCTCG	395	4
Mapk1	Map kinase 1	*Pst*	HM535614.1	Map1351S:GTCGGTCGGGTGTATCCT Map1683A:GGTTCATCTTCGGGGTCA	332	4
Cs	Citrate synthase	*Pst*	AJIL010000431	Cs1295S:TGGACTCGCCAACCAGGAA Cs1589A: TGGGTCGGCTGCCACTTCTG	295	4
Hsp	Hheat sock protein 90kDa (hsp90)	*Pst*	AJIL01000023.1	Hsp702S: CTACTCTGCCTACCTCGTT Hsp1103A: GATCTTGGCTTCCTTACTTT	400	6
Uba	Ubiquitin-activating enzyme E1	*Pst*	AJIL01000094.1	Uba1715S: ACCCAAACCACGGAACCC Uba2088A: TCGCTCCAGCACCAACTA	374	4
Ubc	Uubiquitin-conjugating enzymeE2	*Pst*	AJIL01000007.1	Ubc279S: TTTGCGAATGGAGTATGG Ubc581A:GAGGGACTGACCTTTGAC	303	7

Note: *Pgt* means the abbreviation of *Puccinia* graminis f. sp. *tritici*.

Li et al. (2014; 2018) developed SNP markers for seven house-keeping genes of *Pst* (Table 1.8). Using these primers on the populations of *Pst*, revealed that both Yunnan and Gansu population of *Pst* had a high genetic diversity, and compared with Gansu population, Yunnan population has ancestral haplotype and higher mutation rate than the Gansu population, but the later has more recombination events. There was a massive gene flow between Yunnan and Gansu. And the direction of gene flow verified by the trajectories of upper air flow, mainly from Yunnan to Gansu. Accordingly, it suggested Yunnan as the source population for *Pst* in China, while Gansu may serve as a secondary source (Li et al., 2021). Combined with the research done by Ali et al. (2014) which suggested that the extended Himalayan regions including China as the origin center for *Pst* worldwide, the study was based on the samples from South Gansu, but considering the high diversity in Yunnan, it could also be likely one of the centers of diversity of *Pst* and potential source for new variants at the worldwide level.

1.10 Genomic Studies on Pathogenicity

Genomic analysis of the *Pst* isolate CYR32 (with 110-Mb draft sequence data) revealed *Pst* has a more diverse gene composition and more genes encoding secreted proteins (Zheng et al., 2013). Re-sequencing analysis indicated significant genetic variation among six isolates collected from different continents. Approximately 35% of SNPs were in the coding sequence regions, and half of them were non-synonymous. High genetic diversity in *Pst* suggested that sexual reproduction plays an important role in the origin of different regional races (Zheng et al., 2013).

Many pathogenicity genes such as *PsMAPK1, PsCaMKL1, PsCNA1* and *PsCNB1,* etc. has been cloned and functionally characterized, their expressions are induced at early infection stages (Liang et al., 2009; Zhang et al., 2012; Jiao et al., 2017; Qi et al., 2018), some are during germ tube elongation like *Pscamk* (Qin et al., 2014), some during haustorium formation like *PsMAPK1*, and some regulate *Pst* growth and development (Guo et al., 2011). For example, the *PsCon1* plays a role in formation or survival of *Pst* urediniospores (Guo et al., 2013), and the *PsSRPKL* is responsible for fungal growth and responses to environmental stresses, contributing to *Pst* virulence in wheat (Cheng et al., 2014). Small RNAs (sRNAs) contributes to impairing host immune responses, like *Pst*-milR1 (Wang et al., 2017).

1.11 Effector Proteome

Identification and functional characterization of *Pst* effectors promise to develop effective and sustainable strategies for controlling the disease. It has been established that effectors, the proteins secreted by the pathogen, play a key role in wheat-*Pst* interaction. Recent advances in genomics and bioinformatics have allowed identification of a large repertoire of candidate effectors, while the evolving high-throughput tools have continued to assist in their functional characterization. The repertoires of effectors have become an important resource for better understanding of effector biology of pathosystems and resistance breeding. In recent years, a significant progress has been made in the field of *Pst* effector biology (Prasad et al., 2019).

Pst secretes numerous effectors to modulate host defense systems. It was reported that there were approximately 2,092 genes of *Pst* encoding secreted proteins (SPs). A substantial diversity of the *Pst* secretome and the lack of sequence similarity or conservation in fungal avirulence or effector genes existed (Zheng et al., 2013). Several reports showed that effector protein genes either form *Pst* haustoria or are expressed in haustoria (Dong et al., 2011; Chen et al., 2019; Xu et al., 2019; Xu et al., 2020). The effector expression is highly induced during the early infection stage in wheat (Zhao et al., 2018; Qi et al., 2019) or during pycniospore formation on the alternate host barberry which plays important roles in the regulation of mating and pathogenesis of *Pst* (Zhu et al., 2018).

More research is expected on *Pst* effectors, to better understand the function of these effectors in wheat-*Pst* interaction, and to devise novel strategies for resistance breeding and disease management.

1.12 Integrated Management of Wheat Yellow Rust

An integrated yellow rust management approach is based on the combination of resistant varieties, cultural practices (such as eliminating volunteer wheat and susceptible barberries, adjusting sowing date, etc.), pathogen survey (including resistance monitoring and races identification), government programs (covering disease surveillance, early-warning system and control unified), and timely chemical control (Fig. 1.3). This is based on various studies aiming at the reducing disease level (Chen et al., 2013; Chen and Kang, 2017; Carmona et al., 2020).

As there is no cultivar resistant to all *Pst* races, the use of fungicides has become one of the most important practices for control the yellow rust nationwide. Seed treatments and foliar spray with fungicides are the major methods for chemical control of yellow rust. Three key techniques are recommended at three key stages: (1) Seed treatment with fungicides at sowing stage; (2) Monitoring with fungicides, controlling when disease spots occur at jointing stage; (3) Spray once to resolve several problems, controlling unified in a large area at heading stage. Which means using a mixture of pesticides, fungicides, plant growth regulator, foliar fertilizer and micro-element fertilizer to manage diseases and pests, dry hot wind, etc., all to achieve a sustainable yield.

Fungicide seed treatments is beneficial to control yellow rust especially in hotspot regions where highly susceptible varieties are grown, which will protect the early leaves from the *Pst*

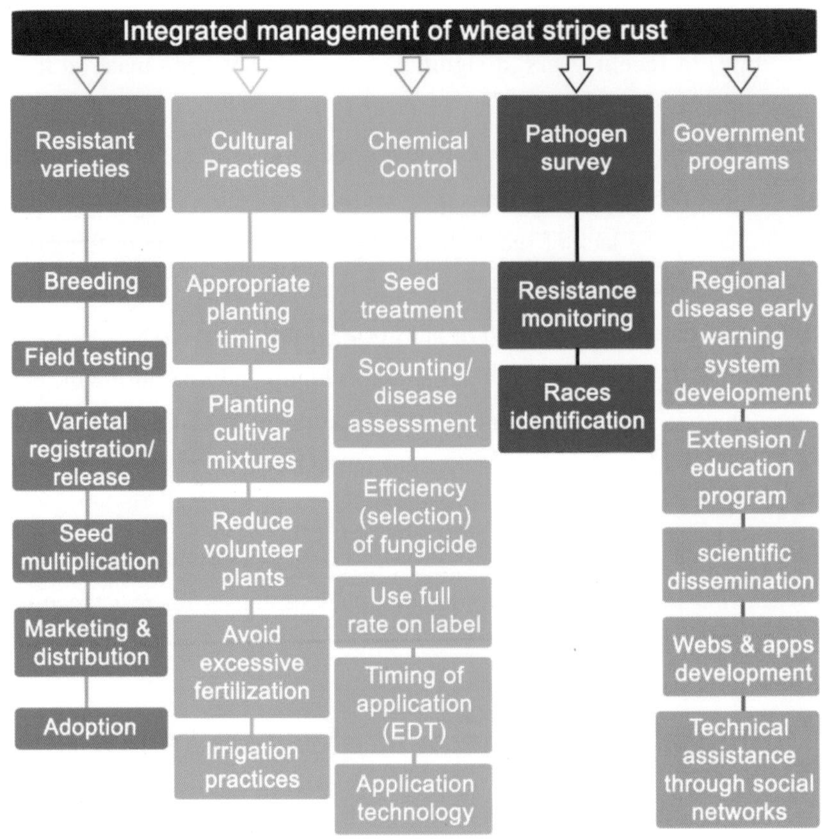

Fig. 1.3 Integrated management and prevention of wheat yellow rust scheme (Carmona et al., 2020)

spores. For foliar spray, in order to avoid unnecessary applications, and minimize negative environmental and social impact, the optimal fungicide application timing should be used. Generally, when the incidence of yellow rust reach to 1% in the field, the foliar fungicide application should be used once or twice, thus it can control the disease well and avoid it spread widely.

Fungicide resistance monitoring programs should be a part of the yellow rust integrated management. Table 1.9 showed the fungicides currently used for control of yellow rust in China.

Chemical control may cause environmental problems, thus the search for new biocontrol methods is obviously necessary. Recent research indicated that some endophytic bacteria had the biocontrol potential and can be used to manage yellow rust disease instead of fungicides (Kiani et al., 2021), more bio-materials and new techniques will be developed to control wheat yellow rust in the future.

Table 1. 9 Fungicides currently used for control of yellow rust in China

Trade mark	Active ingredients	Chemical group	Mode of action	FRAC group	Main action type	Plant organ	Reference
Bayleton®	Triadimenol	DMI	C14-demethylase in sterol biosynthesis	3	Curative	Foliar	Wan et al., 2007
Tilt®	Propiconazole	DMI	C14-demethylase in sterol biosynthesis	3	Curative	Foliar	Chen, 2007
Stratego®	Propiconazole + trifloxystrobin	DMI + QoI	C14-demethylase in sterol biosynthesis + cytochrome bc1 (ubiquinol oxidase)	3 + 11	Curative + preventive	Foliar	Chen, 2007
Prosaro®	Prothioconazole + tebuconazole	DMI	C14-demethylase in sterol biosynthesis	3	Curative	Foliar	Chen, 2007
Quilt®	Propiconazole + azoxystrobin	DMI + QoI	C14-demethylase in sterol biosynthesis + cytochrome bc1 (ubiquinol oxidase)	3 + 11	Curative + preventive	Foliar	Chen, 2007
Evito®	Fluoxastrobin	QoI	Cytochrome bc1 (ubiquinol oxidase)	11	Preventive	Foliar	Chen, 2007
Quadris®	Azoxystrobin	QoI	Cytochrome bc1 (ubiquinol oxidase)	11	Preventive	Foliar	Chen, 2007

Note: The table cited from Carmona et al. (2020), DMI is the abbreviation of demethylation inhibitor, and QoI is the abbreviation of quinone outside inhibitor.

1.13 Future Avenues to Work on

Continued forecasting regarding predominant race of *Pst*, studying the pathogenicity and variation of *Pst*, and use of new techniques to modify and change key genes to slow down the variation of pathogen are the promising areas where future work should be done. Combining gene mapping and postulation to understand the distribution of Rgenes and its allelic relationship, fine mapping of these Rgenes, should be encouraged to develop durable resistance by pyramiding major resistant gene or APR genes. Finally, strengthening the related research on alternate hosts, breeding new resistant cultivars, as well as developing new bio-fungicides will help to adopt an integrated disease management strategy.

References

Ahmed SM, Liu P, Xue Q, et al., 2017. *TaDIR1-2*, a Wheat Ortholog of Lipid Transfer Protein AtDIR1 Contributes to Negative Regulation of Wheat Resistance against *Puccinia striiformis* f. sp. *tritici*. *Front Plant Sci*, 8: 521. DOI: 10.3389/fpls.2017.00521.

Alam MA, Li H, Hossain A, et al., 2021. Genetic Diversity of Wheat Stripe Rust Fungus *Puccinia striiformis* f.

sp. *tritici* in Yunnan, China. Plants (Basel), 10(8):1735. DOI: 10.3390/plants10081735.

Al-Attala MN, Wang X, Abou-Attia MA, et al., 2014. A Novel *TaMYB4* Transcription Factor Involved In the Defence Response Against *Puccinia striiformis* f. sp. *tritici* and Abiotic Stresses. *Plant Mol Biol*, 84(4-5):589-603. DOI: 10.1007/s11103-013-0156-7.

Ali S, Gladieux P, Leconte M, et al., 2014. Origin, Migration Routes and Worldwide Population Genetic Structure of the Wheat Yellow Rust Pathogen *Puccinia striiformis* f. sp. *tritici*. *PLoS Pathogen*, 10 (1): e1003903. DOI:10.1371/journal.ppat.1003903.

Ali S, Leconte M, Walker A-S, et al., 2010. Reduction in the Sex Ability of Worldwide Clonal Populations of *Puccinia striiformis* f. sp. *tritici*. *Fungal Genetics and Biology,* 47:828-838.

Ali S, Rodriguez-Algaba J, Thach T, et al., 2017. Yellow Rust Epidemics Worldwide were Caused by Pathogen Races from Divergent Genetic Lineages. *Frontiers in Plant Science* , 8:1058.

Ali S, Sharma S, Leconte M, et al., 2018. Low Pathotype Diversity in a Recombinant *Puccinia striiformis* Population Through Convergent Selection at the Eastern Part of Himalayan Centre of Diversity (Nepal). *Plant Pathology,* 67:810-820.

Bansal U, Hayden M, Keller B, et al., 2009. Relationship Between Wheat Rust Resistance Genes Yr1 and Sr48 and a Microsatellite Marker. *Plant Pathology*, 58(6):1039-1043. DOI: 10.1111/j.1365-3059.2009.02144.x.

Carmona M, Sautua F, Pérez-Hérnandez O, et al., 2020. Role of Fungicide Applications on the Integrated Management of Wheat Yellow Rust. *Front Plant Sci.*, 11:733. DOI: 10.3389/fpls.2020.00733.

Chen CQ, 2008. Molecular Population Genetic Structure of *Puccinia striiformis* f. sp. *tritici* in China. Ph.D. Dissertation, Northwest Agriculture & Forestry University, China.

Chen WQ, Kang ZS, Ma ZH, et al., 2013. Integrated Management of Wheat Yellow Rust Caused by *Puccinia striiformis* f. sp. *tritici* in China. *Scientia Agricultura Sinica,* 46 (20):4254-4262. (in Chinese with English abstract).

Chen WQ, Wu LR, Liu TG, et al., 2009. Race Dynamics, Diversity, and Virulence Evolution in *Puccinia striiformis* f. sp. *tritici*, the Causal Agent of Wheat Yellow Rust in China from 2003 to 2007. *Plant Disease*, 93: 1093-1101.

Chen XM, Kang ZS (eds.), 2017. *Yellow Rust*, Dordrecht: Springer. DOI: 10.1007/978-94-024-1111-9_7.

Chen XM, Soria MA, Yan GP, et al., 2003. Development of Sequence Tagged Site and Cleaved Amplified Polymorphic Sequence Markers for Wheat Yellow Rust Resistance Gene *Yr5*. *Crop Science*, 43: 2058-2064.

Chen Y, Mao H, Wu N, et al., 2019. Effects of Yellow Rust Infection on the Levels of Redox Balance and Photosynthetic Capacities in Wheat. *International Journal of Molecular Sciences*, 21(1):268. DOI: 10.3390/ijms21010268.

Chen ZJ, Wang T, Tang CL, et al., 2019. Funtional Analysis of *Puccinia striiformis* f. sp. *tritici* Effector Hasp58 Inhibits Plant Immunity. *Journal of Triticeae Crops,* 39(2):239-246. (in Chinese with English abstract).

Cheng B, Gao X, Cao N, et al., 2020. Genome-wide Association Analysis of Yellow Rust Resistance Loci in

Wheat Accessions from Southwestern China. *J Appl Genet*, 61(1):37-50. DOI: 10.1007/s13353-019-00533-8.

Cheng L, Yang ZJ, Li GR, et al., 2008. Isolation of a New Repetitive DNA Sequence From Secale Afticanum Enables Targeting of Secale Chromation in Wheat Background. *Euphytica,* 159(1-2):251.

Cheng P, Xu LS, Wang MN, et al., 2014. Molecular Mapping of Genes *Yr64* and *Yr65* for Yellow Rust Resistance in Hexaploid Derivatives of Durum Wheat Accession PI 331260 and PI 480016. *Theor Appl Genet*, 127:2267–77.

Cheng YL, Wang XJ, Yao JN, et al., 2014. Characterization of Protein Kinase PsSRPKL, a Novel Pathogenicity Factor in the Wheat Yellow Rust Fungus. *Environmental Microbiology,* DOI: 10.1111/1462-2920.12719.

Dong N, Hu H, Hu TZ, et al., 2019. Molecular Detection and Distribution of Yellow Rust Resistance Genes *Yr5*, *Yr10* and *Yr18* among 348 Wheat Germplasms. *Acta Agriculturae Boreali-occidentalis Sinica*, 28(12):1960-1968. DOI: 10.7606/j.issn.1004-1389.2019.12.007.

Dong YL, Yin CT, Hulbert S, et al., 2011. Cloning and Expression Analysis of Three Secreted Protein Genes from Wheat Yellow Rust Fungus *Puccinia striiformis* f. sp. *tritici*. *World Journal of Microbiology & Biotechnology,* 27(5):1261-1265.

Duan X, Tellier A, Wan A, et al., 2010. *Puccinia striiformis* f. sp. *tritici* Presents High Diversity and Recombination in the Over-summering Zone of Gansu, China. *Mycologia,* 102(1):44-53.

Fang CT, 1944. Physiologic Specialization of *Puccinia glumarum* Erikss and Henn. in China. *Phytopathology*, 34:1020-1024.

Feng H, Wang X, Sun Y, et al., 2011. Cloning and Characterization of a Calcium Binding EF-hand Protein Gene *TaCab1* from Wheat and its Expression in Response to *Puccinia striiformis* f. sp. *tritici* and Abiotic Stresses. *Mol Biol Rep*, 38(6): 3857-3866.

Francis H, Leitch A, Koebner R, 1995. Conversion of a RAPD-generated PCR Product, Containing a Novel Dispersed Repetitive Element, into a Fast and Robust Assay for the Presence of Rye Chromatin in Wheat. *Theoretical and Applied Genetics,* 90(5): 636-642.

Guo J, Duan YH, Zhang JS, et al., 2013. A Conidiation-related Gene is Highly Expressed at the Resting Urediospore Stage in *Puccinia striiformis* f. sp. *tritici*. *Journal of Basic Microbiology,* 53(8): 695-702.

Guo J, Dai X, Xu J, et al., 2011. Molecular Characterization of a Fus3/Kss1 type *MAPK* from *Puccinia striiformis* f. sp. *tritici*, *PsMAPK1*. *PLoS One,* 6(7):e21895. DOI:10.1371/journal.pone.0021895.

Han DJ, Kang ZS, 2018. Current Status and Future Strategy in Breeding Wheat for Resistance to Yellow Rust in China. *Plant Protection,* 44(5):1-2. (in Chinese with English abstract).

He ZH, Lan CX, Chen XM, et al., 2011. Progress and Perspective in Research of Adult-Plant Resistance to Yellow Rust and Powdery Mildew in Wheat. *Scientia Agricultura Sinica,* 44(11):2193-2215. (in Chinese with English abstract).

He ZH, Zhuang QS, Cheng SH, et al., 2018. Wheat Production and Technology Improvement in China. *Journal of Agriculture,* 8(1):99-106. (in Chinese with English abstract).

Helguera M, Khan IA, Kolmer J, et al., 2003. PCR Assays for the *Lr37-Yr17-Sr38* Cluster of Rust Resistance Genes and Their Use to Develop is Isogenic Hard Red Spring Wheat Lines. *Crop Science*, 43(5):1842.

http://www.china.org.cn/e-china/geography/index.htm.

http://www.stats.gov.cn/tjsj./ndsj/.

https://www.ars.usda.gov/midwest-area/stpaul/cereal-disease-lab/docs/resistance-genes/resistance-genes.

Hu XP, Wang BT, Kang ZS, 2014. Research Progress on Virulence Variation of *Puccinia striiformis* f. sp. *tritici* in China. *Journal Tritceae Crop,* 34(5):709-716. (in Chinese with English abstract).

Huai B, Yang Q, Qian Y, et al., 2019. ABA-Induced Sugar Transporter *TaSTP6* Promotes Wheat Susceptibility to Yellow Rust. *Plant Physiology,* 181(3):1328-1343. DOI: 10.1104/pp.19.00632.

Huang S, Liu S, Zhang Y, et al., 2020. Genome-wide Wheat 55K SNP-based Mapping of Yellow Rust Resistance Loci in Wheat Cultivar Shaannong 33 and Their Alleles Frequencies in Current Chinese Wheat Cultivars and Breeding Lines. *Plant Dis,* DOI: 10.1094/PDIS-07-20-1516-RE.

Jiang Z, Ge S, Xing L, et al., 2003. *RLP1.1*, a Novel Wheat Receptor-like Protein Gene, is Involved in the Defence Response Against *Puccinia striiformis* f. sp. *tritici*. *J Exp Bot,* 64(12):3735-3746. DOI:10.1093/jxb/ert206.

Jiao M, Yu D, Tan C, et al., 2017. Basidiomycete-specific *PsCaMKL1* Encoding a CaMK-like Protein Kinase is Required for Full Virulence of *Puccinia striiformis* f. sp. *tritici*. *Environ Microbiol,* 19(10):4177-4189. DOI:10.1111/1462-2920.13881.

Jin Y, Szabo LJ, Carson M, 2010. Century-old Mystery of *Puccinia striiformis* Life History Solved with the Identification of *Berberis* as an Alternate Host. *Phytopathology,* 100(5):432-435.

Kang ZS, Wang XJ, Zhao J, et al., 2015. Advances in Research of Pathogenicity and Virulence Variation of the Wheat Yellow Rust Fungus *Puccinia striiformis* f. sp. *tritici*. *Scientia Agricultura Sinica,* 48(17):3439-3453. (in Chinese with English abstract).

Kiani T, Mehboob F, Hyder MZ, et al., 2021. Control of Yellow Rust of Wheat Using Indigenous Endophytic Bacteria at Seedling and Adult Plant Stage. *Sci Rep,* 11(1):14473. DOI: 10.1038/s41598-021-93939-6.

Klymiuk V, Yaniv E, Huang L, et al., 2018. Cloning of the Wheat *Yr15* Resistance Gene Sheds Light on the Plant Tandem Kinase-pseudokinase Family. *Nature Communications*, (9):3747.

Lagudah ES, Krattinger SG, Herrera-Foessel S, et al., 2009. Gene-specific Markers for the Wheat Gene *Lr34/Yr18/Pm38* Which Confers Resistance to Multiple Fungal Pathogens. *Theoretical and Applied Genetics,* 119:889-898.

Lagudah ES, McFadden H, Singh RP, et al., 2006. Molecular Genetic Characterization of the *Lr34/Yr18* Slow Rusting Resistance Gene Region in Wheat. *Theoretica and Applied Genetics,* 114:21-30.

Lan C, Liang S, Zhou X, et al., 2010. Identification of Genomic Regions Controlling Adult-plant Yellow Rust Resistance in Chinese Landrace Pingyuan 50 Through Bulked Segregant Analysis. *Phytopathology,* 100(4):313-8. DOI: 10.1094/PHYTO-100-4-0313.

Lesleyr RM, Dipak S, Kimberlee K, et al., 2009. Linkage Maps of Wheat Yellow Rust Resistance Genes *Yr5* and *Yr15* for Use in Marker-assisted Selection. *Crop Science*, 49 (5):1786-1790.

Li FQ, Han DJ, Wei GR, et al., 2008. Molecular Detection of Yellow Rust Resistant Genes in 126 Winter Wheat Varieties from the Huanghuai Wheat Region. *Scientia Agricultura Sinica,* 41(10):3060-3069. (in Chinese with English abstract).

Li J, Jiang Y, Yao F, et al., 2020. Genome-Wide Association Study Reveals the Genetic Architecture of Yellow Rust Resistance at the Adult Plant Stage in Chinese Endemic Wheat. *Front Plant Sci,* 11:625. DOI: 10.3389/fpls.2020.00625.

Li MJ, Alam MA, Li HX, et al., 2018. Development and Polymorphic Loci Research of SNP Primers from House-keeping Genes of *Puccinia striiformis* f. sp. *tritici*. *Molecular Plant Breeding*, 16(5):1539-1544. (in Chinese with English abstract).

Li MJ, Chen WQ, Duan XY, et al., 2014. First Report of SNP Primers of Three House-keeping Genes of *Puccinia striiformis* f. sp. *tritici*. *Acta Phytopathologica Sinica,* 44(5):536-541.

Li MJ, Chen XM, Wan AM, et al., 2018. Virulence Characterization of Yellow Rust Pathogen *Puccinia striiformis* f. sp. *tritici* Population to 18 Near-isogenic Lines Resistant to Wheat Yellow Rust in Yunnan Province. *Journal of Plant Protection,* 45(1):75-82. (in Chinese with English abstract).

Li MJ, Feng J, Cao SQ, et al., 2011. Postulation of Seedlings Resistance Genes to Yellow Rust in Commercial Wheat Cultivars from Yunnan Province in China. *Agricultural Sciences in China*, 10 (11), 1723-1731.

Li MJ, Zhang YH, Chen WQ, et al., 2021. Evidence for Yunnan as the Major Origin Center of the Dominant Wheat Fungal Pathogen *Puccinia striiformis* f. sp. *tritici*. *Australasian Plant Pathology,* 50(2):241-252. DOI:10.1007/s13313-020-00770-0.

Li SN, Chen W, Ma XY, et al., 2021. Identification of Eight *Berberis* Species from the Yunnan-Guizhou Plateau as Aecial Hosts for *Puccinia striiformis* f. sp. *tritici*, the Wheat Stripe Rust Pathogen. *Journal of Integrative Agriculture,* 20(6):1563-1569.

Li ZQ, Zeng SM, 2002. Wheat Rust in China. Beijing: China Agriculture Press. (in Chinese).

Liang XF, Liu B, Zhu L, et al., 2009. Cloning and Expression of a Class Ⅱ Chitin Synthase Gene *PstChs II* from the Rust Fungus *Puccinia striiformis*. *Acta Microbiologica Sinica,* 49(12):1621-1627. (in Chinese with English abstract).

Liu B, Liu TG, Zhang ZY, et al., 2017. Discovery and Pathogenicity of CYR34, a New Race of *Puccinia striiformis* f. sp. *tritici* in China. *Acta Phytopathologica Sinica,* 47(5):681-687. (in Chinese with English abstract).

Liu LJ, Wang ZL, Xi YJ, et al., 2008. Detection of Yellow Rust Resistant Gene *Yr26* with SSR Markers in Wheat Cultivars of Huanghuai Region. *Acta Bot. Boreal.-Occident.Sin*, 28(7):1308-1312.

Liu W, Frick M, HUEL R, et al., 2014. Yellow Rust Resistance Gene *Yr10* Encodes an Evolutionary-conserved and Unique CC-NBS-LRR Sequence in Wheat. *Molecular Plant*, 7(12):1750.

Liu YP, Cao SH, Wang XP, et al., 2005. Molecular Mapping of Yellow Rust Resistance Gene *Yr24* in Wheat. *Acta Phytopathologica Sinica*, 35(5):478-480. (in Chinese with English abstract).

Lu NH, Zhan GM, Wang JF, et al., 2009. Molecular Evidence of Somatic Genetic Recombination of *Puccinia striiformis* f. sp. *tritici* in China. *Acta Phytopathologica Sinica,* 39(6):561-568. (in Chinese with English abstract).

Lu Y, Lan C, Liang S, et al., 2009. QTL Mapping for Adult-plant Resistance to Yellow Rust in Italian Common Wheat Cultivars Libellula and Strampelli. *Theor Appl Genet*, 119(8):1349-59. DOI: 10.1007/s00122-009-1139-6.

Lv XH, Feng J, Lin RM, et al., 2013. Analysis of Resistance Genes and Identification at Adult Stage to Yellow Rust in 30 New Wheat Lines, *Acta Phytopathologica Sinica*, 43(3):323-327. (in Chinese with English abstract).

Ma ZH, Shi SD, Wang HG, et al., 2005. Climate-based Regional Classification for Oversummering and Overwintering of *Puccinia striiformis* in China with GIS and Geostatistics. *Journal of Northwest Sci-Tech University of Agriculture and Forestry (Natural Science Edition)*, (S1):11-13. (in Chinese with English abstract).

Mago R, Mian H, Lawrence GJ, et al., 2005. High-resolution Mapping and Mutaion Analysis Separate the Rust Resistance Gene *Sr31*, *Lr26* and *Yr9* on the Short Arm of Rye Chromosome 1. *Theoretical and Applied Genetics,* 112(1):41.

Mamun MA, Tang C, Sun Y, et al., 2018. Wheat Gene TaATG8j Contributes to Yellow Rust Resistance. *Internationa Journal of Molecular Sciences,* 19(6):1666. DOI: 10.3390/ijms19061666.

Marchal C, Zhang J, Zhang P, et al., 2018. BED-domain-containing Immune Receptors Confer Diverse Resistance Spectra to Yellow Rust. *Nat Plants,* 4(9):662-668. DOI:10.1038/s41477-018-0236-4.

McIntosh RA, Dubcovsky J, Rogers J, et al., 2018. Catalogue of Gene Symbols for Wheat: 2017 Supplement[2018-08-01]. Available online: http://shigen.nig.ac.jp/wheat/komugi/genes/macgene/supplement2017.pdf.

McIntosh RA, Wellings CR, Park RF, 1995. Wheat Rust: an Atlas of Resistance Genes. Melbourne: CSIRO Publishing.

Mehmood S, Sajid M, Zhao J, et al., 2020. Alternate Hosts of *Puccinia striiformis* f. sp. *tritici* and Their Role. *Pathogens,* 9(6):434. DOI: 10.3390/pathogens9060434.

Mittermeier RA, 1988. Primate Diversity and the Tropical Forest: Case Studies from Brazil and Madagascar and the Importance of the Megadiversity Countries. in Biodiversity (ed. Wilson EO).Washington, D. C.: National Academy Press.

Mu J, Huang S, Liu S, et al., 2019. Genetic Architecture of Wheat Yellow Rust Resistance Revealed by Combining QTL Mapping Using SNP-based Genetic Maps and Bulked Segregant Analysis. *Theor Appl Genet*, 132(2):443-455. DOI: 10.1007/s00122-018-3231-2.

Mu J, Wu J, Liu S, et al., 2019. Genome-Wide Linkage Mapping Reveals Yellow Rust Resistance in Common Wheat (*Triticum aestivum*) Xinong1376. *Plant Dis,* 103(11):2742-2750. DOI: 10.1094/PDIS-12-18-2264-RE.

Murphy LR, Santra D, Kidwell K, et al., 2009. Linkage Maps of Wheat Yellow Rust Resistance Genes and for Use in Marker-assisted Selection. *Crop Science,* 49(5):1786-1790.

Myers N, Mittermeier RA, Mittermeier CG, et al., 2000. Biodiversity Hotspots for Conservation Priorities. *Nature,* 403(6772):853-8. DOI: 10.1038/35002501.

Peng JH, Fahima T, Roeder MS, et al., 2000. High-density Molecular Map of Chromosome Region Harboring Yellow-rust Resistance Genes *YrH52* and *Yr15* Derived from Wild Emmer Wheat. *Triticum dicoccoides, Genetica,* 109(3):199-210.

Prasad P, Savadi S, Bhardwaj SC, et al., 2019. Rust Pathogen Effectors: Perspectives in Resistance breeding. *Planta,* 250(1):1-22. DOI: 10.1007/s00425-019-03167-6.

Qi T, Guo J, Liu P, et al., 2019. Yellow Rust Effector *PstGSRE1* Disrupts Nuclear Localization of ROS-Promoting Transcription Factor *TaLOL2* to Defeat ROS-Induced Defense in Wheat. *Mol Plant,* 12(12):1624-1638. DOI: 10.1016/j.molp.2019.09.010.

Qi T, Zhu X, Tan C, et al., 2018. Host-induced Gene Silencing of an Important Pathogenicity Factor *PsCPK1* in *Puccinia striiformis* f. sp. *tritici* Enhances Resistance of Wheat Yellow Rust. *Plant Biotechnol J,* 16(3):797-807.

Qin J, Huang C, He F, et al., 2014. Function of a Calcium-dependent Protein Kinase Gene Pscamk in *Puccinia striiformis* f. sp. *tritici*. *Actc Microbiologica Sinica,* 54(11):1296-1303. (in Chinese with English abstract).

Ren Y, Li Z, He Z, et al., 2012. QTL Mapping of Adult-plant Resistances to Yellow Rust and Leaf Rust in Chinese Wheat Cultivar Bainong 64. *Theor Appl Genet*, 125(6):1253-62. DOI: 10.1007/s00122-012-1910-y.

Roder MS, Korzun V, Wendekhake K, et al., 1998. A Microsatellite Map of Wheat. *Genetics*, 149: 2007-2023.

Rodriguez-Algaba J, Hovmøller MS, Villegas D, et al., 2021. Two Indigenous *Berberis* Species from Spain Were Confirmed as Alternate Hosts of the Yellow Rust Fungus *Puccinia striiformis* f. sp. *tritici*. *Plant Dis*, 105(9):2281-2285. DOI:10.1094/PDIS-02-21-0269-SC.

Rosewarne GM, Herrera-Foessel SA, Singh RP, et al., 2013. Quantitative Trait Loci of Yellow Rust Resistance in Wheat. *Theor Appl Genet*, 126(10):2427-49. DOI: 10.1007/s00122-013-2159-9.

Shan WX, Chen SY, Kang ZS, et al., 1998. Genetic Diversity in *Puccinia striiformis* Westend. f. sp. *tritici* Revealed by Pathogen Genome-specific Repetitive Sequence. *Canadian Journal of Botany,* 76(4):587-595.

Shang HS, Jing JX, Li ZQ, 1994. Mutations Induced by Ultraviolet Radiation Affecting Virulence in *Puccinia striiformis*. *Acta Phytopathologica Sinica,* 24(4):47-351. (in Chinese with English abstract).

Shao YT, Niu YC, Zhu LH, et al., 2001. Mapping the Yellow Rust Resistance Gene *Yr10* with AFLP Marker. *Chinese Science Bulletin,* 46(8):669-672. (in Chinese with English abstract).

Sinha P, Chen X, 2021. Potential Infection Risks of the Wheat Yellow Rust and Stem Rust Pathogens on Barberry in Asia and Southeastern Europe. *Plants* (Basel), 10(5):957. DOI: 10.3390/plants10050957.

Smith PH, Hadfield J, Hart NJ, et al., 2007. STS Markers for the Wheat Yellow Rust Resistance Gene *Yr5* Suggest a NBS-LRR-type Resistance Gene Cluster. *Genome*, 50(3):260.

Tao F, Wang J, Guo Z, et al., 2018. Transcriptomic Analysis Reveal the Molecular Mechanisms of Wheat Higher-Temperature Seedling-Plant Resistance to *Puccinia striiformis* f. sp. *tritici*. *Front Plant Sci*, 9:240. DOI: 10.3389/fpls.2018.00240.

Wan AM, Chen XM, He ZH, 2007. Wheat Yellow Rust in China. *Australian Journal Agriculture Research*, 58:605-619.

Wan AM, Zhao ZH, Chen XM, et al., 2004. Wheat Yellow Rust Epidemics and Virulence of *Puccinia striiformis* f. sp. *tritici* in China in 2002. *Plant Disease*, 88:896-904.

Wang B, Sun Y, Song N, et al., 2017. *Puccinia striiformis* f. sp. *tritici* microRNA-like RNA 1 (Pst-milR1), an Important Pathogenicity Factor of *Pst*, Impairs Wheat Resistance to *Pst* by Suppressing the Wheat Pathogenesis-related 2 gene. *New Phytol*, 215(1):338-350. DOI:10.1111/nph.14577.

Wang CM, Zhang YP, Han DJ, et al., 2008. SSR and STS Markers for Wheat Yellow Rust Resistance Gene *Yr26*. *Euphytica*, 159:359-366.

Wang H, Zou S, Li Y, et al., 2020. An Ankyrin-repeat and WRKY-domain containing Immune Receptor Confers Yellow Rust Resistance in Wheat. *Nat Commun*, 11(1):1353. DOI: 10.1038/s41467-020-15139-6.

Wang J, Tian W, Tao F, et al., 2020. TaRPM1 Positively Regulates Wheat High-Temperature Seedling-Plant Resistance to *Puccinia striiformis* f. sp. *tritici*. *Front Plant Sci*, 10:1679. DOI: 10.3389/fpls.2019.01679. eCollection 2019.

Wang J, Wang J, Shang H, et al., 2019. *TaXa21*, a Leucine-Rich Repeat Receptor-Like Kinase Gene Associated with *TaWRKY76* and *TaWRKY62*, Plays Positive Roles in Wheat High-Temperature Seedling Plant Resistance to *Puccinia striiformis* f. sp. *tritici*. *Mol Plant Microbe Interact*, 32(11):1526-1535. DOI: 10.1094/MPMI-05-19-0137-R.

Wang JF, Chen CQ, Lu NH, et al., 2010. SSR Analysis of Population Genetic Diversity of *Puccinia striiformis* f. sp. *tritici* in Sichuan Province, China. *Mycosystema*, 29(2):206-213. (in Chinese with Englsih abstract).

Wang LF, Ma JX, Zhou RH, et al., 2002. Molecular Tagging of the Yellow Rust Resistance Gene *Yr10* in Common Wheat, P.I.178383 (*Triticum aestivum* L.). *Euphytica*, 124:71-73.

Wang M, Chen X, 2013. First report of Oregon Grape (*Mahonia aquifolium*) as an Alternate Host for the Wheat Yellow Rust Pathogen (*Puccinia striiformis* f. sp. *tritici*) Under Artificial Inoculation. *Plant Disease*, 97(6):839. DOI:10.1094/PDIS-09-12-0864-PDN.

Wang S, Li QP, Wang J, et al., 2019. *YR36*/WKS1-Mediated Phosphorylation of PsbO, an Extrinsic Member of Photosystem II, Inhibits Photosynthesis and Confers Yellow Rust Resistance in Wheat. *Mol Plant*, 12(12):1639-1650.DOI: 10.1016/j.molp.2019.10.005.

Wang X, Wang X, Feng H, et al., 2012. *TaMCA4*, a Novel Wheat Metacaspase Gene Functions in Programmed Cell Death Induced by the Fungal Pathogen *Puccinia striiformis* f. sp. *tritici*. *Mol Plant Microbe Interact*,

25(6):755-64. DOI: 10.1094/MPMI-11-11-0283-R.

Wang X, Zhang HG, Liu BL, et al., 2011. Molecular Detection of *Yr10*, *Yr15* Genes and 1BL/1RS Translocation in Qinghai Wheat Cultivars. *Acta Botany Boreal.-Occident. Sinica,* 31(1):0057-0063. (in Chinese with English abstract).

Wu J, Huang S, Zeng Q, et al., 2018. Comparative Genome-wide Mapping Versus Extreme Pool-genotyping and Development of Diagnostic SNP Markers Linked to QTL for Adult Plant Resistance to Yellow Rust in Common Wheat. *Theor Appl Genet,* 131(8):1777-1792. DOI: 10.1007/s00122-018-3113-7.

Wu J, Wang Q, Liu S, et al., 2017. Saturation Mapping of a Major Effect QTL for Yellow Rust Resistance on Wheat Chromosome 2B in Cultivar Napo 63 Using SNP Genotyping Arrays. *Front Plant Sci,* 8:653. DOI: 10.3389/fpls.2017.00653.

Wu Y, Wang Y, Yao F, et al., 2021. Molecular Mapping of a Novel Quantitative Trait Locus Conferring Adult Plant Resistance to Yellow Rust in Chinese Wheat Landrace Guangtoumai. *Plant Dis*, PDIS07201544RE. DOI: 10.1094/PDIS-07-20-1544-RE.

Xu MR, Lv XH, Wang FT, et al., 2019. Detection and Postulation of Resistance Genes to Wheat Yellow Rust in 70 Wheat Varieties (Lines), *Journal of Triticeae Crops,* 39(12):1427-1436. (in Chinese with English abstract).

Xu Q, Li B, Xue WB, et al., 2016. Establishment of Resistance Evaluation System for the Wheat Yellow Rust Resistance Genes *Yr1* and *Yr2*, *Journal of Triticeae Crops*, 36(12):1605-1610. DOI: 10.7606/j.issn.1009-1041.2016.12.08. (in Chinese with English abstract).

Xu Q, Tang C, Wang L, et al., 2020. Haustoria - arsenals During the Interaction Between Wheat and *Puccinia striiformis* f. sp. *tritici*. *Mol Plant Pathol,* 21(1):83-94. DOI: 10.1111/mpp.12882.

Xu Q, Tang C, Wang X, et al., 2019. An Effector Protein of the Wheat Yellow Rust Fungus Targets Chloroplasts and Suppresses Chloroplast Function. *Nat Commun*, 10(1):5571. DOI:10.1038/s41467-019-13487-6.

Xu XD, Wang MY, Feng J, et al., 2018. Analysis of the Yellow Rust Resistant Genes in 22 Chinese Wheat Cultivars. *Journal of Plant Protection*, 45(1):37-45. (in Chinese with English abstract).

Xue WB, Xu X, Mu JM, et al., 2014. Evaluation of Yellow Rust Resistance and Genes in Chinese Wheat Varieties. *Journal of Triticeae Crops,* 34(8):1054-1060. (in Chinese with English abstract).

Yang M, Li G, Wan H, et al., 2019. Identification of QTLs for Yellow Rust Resistance in a Recombinant Inbred Line Population. *Int J Mol Sci*, 20(14):3410. DOI: 10.3390/ijms20143410.

Yang WX, Yang FP, Liang D, et al., 2008. Molecular Characterization of Slow-Rusting Genes *Lr34/Yr18* in Chinese Wheat Cultivars. *Acta Agronomica Sinica,* 34(7):1109-1113. (in Chinese with English abstract).

Yang Y, Zhang F, Zhou T, et al., 2021. In Silico Identification of the Full Complement of Subtilase-encoding Genes and Characterization of the Role of *TaSBT1.7* in Resistance Against Yellow Rust in Wheat. *Phytopathology*, 111(2): 398-407.

Yao F, Zhang X, Ye X, et al., 2019. Characterization of Molecular Diversity and Genome-wide Association Study of Yellow Rust Resistance at the Adult Plant Stage in Northern Chinese Wheat Landraces. *BMC Genet,*

20(1):38. DOI: 10.1186/s12863-019-0736-x.

Ye X, Li J, Cheng Y, et al., 2019. Genome-wide Association Study of Resistance to Yellow Rust (*Puccinia striiformis* f. sp. *tritici*) in Sichuan Wheat. *BMC Plant Biol*, 19(1):147. DOI: 10.1186/s12870-019-1764-4.

Ying JS, Boufford DE, Brach AR, 2011. *Flora of China*, 19:714-782. http://www.iplant.cn/foc/pdf/Berberidaceae.pdf.

Yuan FP, Zeng QD, Wu JH, et al., 2018. QTL Mapping and Validation of Adult Plant Resistance to Yellow Rust in Chinese Wheat Landrace Humai 15. *Front Plant Sci*, 9:968. DOI: 10.3389/fpls.2018.00968.

Zeng Q, Han D, Wang Q, et al., 2014. Yellow Rust Resistance and Genes in Chinese Wheat Cultivars and Breeding Lines. *Euphytica,* 196(2):271-284.

Zeng Q, Wu J, Liu S, et al., 2019. Genome-wide Mapping for Yellow Rust Resistance Loci in Common Wheat Cultivar Qinnong 142. *Plant Dis,* 103(3):439-447. DOI: 10.1094/PDIS-05-18-0846-RE.

Zhang H, Guo J, Voegele RT, et al., 2012. Functional Characterization of Calcineurin Homologs *PsCNA1*/*PsCNB1* in *Puccinia striiformis* f. sp. *tritici* Using a Host-induced RNAi System. *PloS One,* 7(11): e49262.

Zhang HY, Wang Z, Ren JD, et al., 2017. A QTL with Major Effect on Reducing Yellow Rust Severity Detected from a Chinese Wheat Landrace. *Plant Dis,* 101(8):1533-1539. DOI: 10.1094/PDIS-08-16-1131-RE.

Zhang P, McIntosh RA, Hoxha S, et al., 2009. Wheat Yellow Rust Resistance Genes *Yr5* and *Yr7* are Allelic. *Theor Appl Genet*, 120:25–9.

Zhang Q, Wang B, Wei J, et al., 2018. *TaNTF2*, a Contributor for Wheat Resistance to the Yellow Rust Pathogen. *Plant Physiology and Biochemistry,* 123:260-267. DOI: 10.1016/j.plaphy.2017.12.020.

Zhang X, Han D, Zeng Q, et al., 2013. Fine Mapping of Wheat Yellow Rust Resistance Gene *Yr26* Based on Collinearity of Wheat with *Brachypodium Distachyon* and Rice. *PloS One,* 8(3):e57885. DOI:10.1371/journal.pone.0057885.

Zhang YW, Liu TG, Liu B, et al., 2014a. Gene Postulation of Yellow Rust Resistance Genes of 75 Chinese Commercial Wheat Cultivars. *Acta Phytophylacica Sinica*, 41(1):45-53. (in Chinese with English abstract).

Zhang YW, Liu B, Liu TG, et al., 2014b. Molecular Detection of *Yr10* and *Yr18* Genes and 1BL/1RS Translocation in Wheat Cultivars. *Plant Protection,* 40(1):54-59. (in Chinese with English abstract).

Zhao J, Wang L, Wang ZY, et al., 2013. Identification of Eighteen *Berberis* Species as Alternate Hosts of *Puccinia striiformis* f. sp. *tritici* and Virulence Variation in the Pathogen Isolates from Natural Infection of Barberry Plants in China. *Phytopathology,* 103(9):927-934.

Zhao M, Wang J, Ji S, et al., 2018. Candidate Effector *Pst_8713* Impairs the Plant Immunity and Contributes to Virulence of *Puccinia striiformis* f. sp. *tritici*. *Front. Plant Sci,* 9:1294. DOI: 10.3389/fpls.2018.01294.

Zheng WM, Huang LL, Huang JQ, et al., 2013. High Genome Heterozygosity and Endemic Genetic Recombination in the Wheat Yellow Rust Fungus. *Nature Communications,* 4:Article number 2673.

Zhu X, Jiao M, Guo J, et al., 2018. A Novel MADS-box Transcription Factor *PstMCM1-1* is Responsible for Full Virulence of *Puccinia striiformis* f. sp. *tritici*. *Environ Microbiol,* 20(4):1452-1463. DOI: 10.1111/1462-

2920.14054.

Acknowledgments

This work was supported by the National Natural Science Foundation of China (31560490), the National Key R&D Program of China (2018YFD0200500), as well as the Major special project of Yunnan Province (202102AE090003, 202102AE090014).

Chapter 2
Wheat Yellow Rust in Nepal

Baidya Nath Mahto
Nepal Agricultural Research Council (NARC), Singhdarbarplaza, Kathmandu, Nepal
Email: bnmahto_7@yahoo.com

Abstract: Agriculture is one of the important components of the economy of Nepal and thus the main contributor to the nation's food security, to which wheat plays an important role. Several obstacles are faced by the wheat crop, which challenge the genetic potential of wheat varieties, thus reducing the overall yield. Among these factors, wheat yellow rust disease, caused by *Puccinia striiformis* is highly important. Previous work has been done in Nepal to cover many aspects, from disease surveillance, field screening to race phenotyping and population genetics of the pathogen. For rust management, several activities had been adopted including development of resistant varieties, early maturing varieties, short duration grain filling varieties, heat tolerant varieties, application of fungicides along with different gene deployment at different locations. The current book chapter aims to describe the status, importance and future management of wheat yellow rust disease in Nepal. It will provide some basic information to devise a sustainable disease management and resistance gene deployment strategy.

Keywords: Wheat Yellow Rust; *Puccinia striiformis* f. sp. *tritici* (*Pst*); Disease Resistance; Diseases Management; Wheat Cultivation; Nepal

2.1 Disease Onset

Wheat (*Triticum aestivum*) is the most extensively grown crop in the world. Globally, wheat is a prime food crop of more than 50% of world population. Wheat is cultivated in 220 million hectares with average productivity of 3.3 t/ha worldwide (FAO, 2014) . It is the vital grain of civilization and food security and ranks first among the three major staple crops as a source of protein in developing countries (CIMMYT, 2013; CIMMYT, 1997-2000). Wheat is the third most important crop after rice and maize in Nepal and cultivated on 0.71 million hectares with total production of 2.19 million tones and the average national productivity is 2.92 t/ha (MoALD, 2020). Wheat is grown all over the country. It contributed 6.98% in AGDP and 2.30% in GDP (MoF, 2014) and also contributes about 26% to national food security. In Nepal, wheat crop was a minor cereal until early 1960s and now it is the third most important staple food crop of Nepal. It occupies 22.6% of total cereal area and contributes 21.3% of the total cereal production in the country (MoAD, 2014/2015). Wheat is widely adapted with its coverage in all the three agro-climatic regions (plain, mid hill and high hill) of the country, ranging from sea level to 4,000 meters above sea level.

As a result of semi-dwarf, high yielding, rusts resistant wheat varieties introduced during mid 1960s and research and development thereafter, wheat production area increased by 7-fold, production by 15-fold and productivity by 2-fold. Wheat is now grown all over the country. Improved wheat varieties cover 97% of the total wheat area in the country (Ann. Rep. 1975 to 2020; Pokharel and Bhatta, 2001).

Impact of wheat research has been magnificent on area, production, productivity and consumption since the establishment of National Wheat Research Program (NWRP) in 1972. This was achieved through the improved technological innovations and their delivery system, increased input availability and use and good coordination among the national and international stakeholders. Most of the wheat growers have adopted improved cultivation practices and varieties with increased use of inputs. Morris et al. (1994) expected that the post-green revolution period from 1990 onwards the first-generation modern varieties (MVs) will be replaced by newer MVs with increased yield potential and superior disease resistance. Area expansion is expected to level off during this period, so the main benefit of the wheat breeding effort will consist of improved genetic potential.

Spring wheat is commercially grown in some high hill areas. The mid-western and far western part of our country appeared to be ideal for wheat cultivation where crop gets longer time to mature (130+ days) as compared to eastern, central and western part where the crop matures comparatively in shorter (115+ days) period (Pokhrel and Bhatta, 2001). Higher latitudes of mid and far western regions probably favor wheat crop because temperature and the altitude are the most important factors for the growth and development of the wheat plant. Hence, the plant takes six months to mature in Kathmandu while more than six months in other part of the country likes Jumla due to higher elevation and low temperature (Pokhrel and Bhatta, 2001). The optimum temperature requirement for vegetative growth of wheat is 16~ 20℃ while for the reproductive growth stage 25~ 30℃. Analysis of daily weather data revealed that abrupt changes in temperature during grain filling period particularly in the plain area causes shriveled grain and lower grain yield.

In 2014-2015, wheat occupied an area of 0.76 million hectares and is estimated that nearly 80% of wheat and 30% of rice area is grown under rice-wheat cropping system. More than 30% rice area is estimated to be fallow in winter, which could be an opportunity for further expansion of wheat area if irrigation facilities are improved. Roughly 12% of wheat area comes under maize-wheat production system particularly in the hills and the rest 5% of wheat area is covered by various production systems such as vegetable-wheat, potato-wheat, summer legume-wheat and others. The sustainable management of the rice-wheat production system along with the development and dissemination of high yielding, disease resistant, climate adaptive and location specific varieties should get top priority to increase wheat productivity. The feeding habit of people has been slowly shifting towards wheat as per capita wheat consumption increased from 17.4 kg in 1972 to 72 kg in 2013. The area, production, and productivity of wheat are in increasing trend. Similarly, the producer wheat prices, annual average flour retail price and per capita consumption are in increasing trend. The gross and net benefit from wheat growing is also positive both in hills and plain regions. The consumption and production of wheat is moving in parallel, however, the consumption is higher than the production. The net trade of wheat is also negative in most of the years.

The low productivity of wheat is the resultant of several factors. Wheat yield is affected by various abiotic and biotic stresses. Drought has been very important in causing yield losses because most of the wheat grown is in rainfed condition in the country. Of the biotic stresses, wheat rusts including yellow rust, leaf rust, and stem rust are definitely the most important diseases that reduce wheat yields in Nepal. Yellow rust caused by *Puccinia striiformis* f. sp. *tritici* (*Pst*) was first recorded in 1964. Yellow rust posed a major threat to wheat production over large areas in Nepal and the virulence of the pathogen have been changing time to time and several rounds of epidemics were encountered in hilly and mountain regions in the country. Yellow rust epidemic was observed during 1985-1987 in hilly districts Sindhupalchowk, Dolkha, Kabre, Dhankuta while another epidemic was also recorded during 1996-1997 in Baglung, Myagdi, Kaski, Parbat and Syanja, hilly districts. Similarly, yellow rust epidemic was experienced during 2004-2006 in Rasuwa, Sindhupalchowk, Dolkha, Kabre, Dang, Ramechhap and Sindhuli, the hilly districts. It is a major disease which poses an important threat to wheat cultivation and occurs mostly on mid & lower hills, river basin and valleys, causing 30% grain yield losses. It causes yield losses up to 80% in Nepal (Ann. Rep. 1975 to 2020). Sometime many local and susceptible varieties are completely destroyed without any grain formation.

Fig. 2.1 Wheat cultivation under diverse agro-ecological zones across Nepal, ranging from high altitude valleys to plain. *Berberis* plants near to wheat field and infection on *Berberis* leaves can be seen in the lower panel of the figure.

2.2 Epidemiology

The yellow rust pathogen *Pst* was shown to complete its life cycle on *Berberis* spp. (*Berberis chinensis*, *B. holstii*, *B. koreana*, *B. vulgaris*) (Jin et al., 2010). The comparison of worldwide samples revealed high sexual production capacity for the Himalayan samples, including Nepal (Ali et al., 2010), which lies in the pathogen center of diversity (Ali et al., 2014). However, the exact role of sexual reproduction to the disease epidemiology is lacking in Nepal. Due to high temperature prevailing in the plains of Nepal during summer months and the following rainy season, wheat rusts in general, may survive while they could over summer in cooler climate of the hills on volunteer wheat tillers, *Berberis* spp., grasses and on the regular summer crop in hills. The rust has been observed at various time points, including the off-season. It was observed on summer planted (June–July planted) wheat at high hills Nigale. Near the China Border at Sindupalchowk the disease was observed in October–November (Mahto et al., 2014). The rust was also found on seedlings of RR21 wheat in early February at other pockets of the district. Mehta (1940) and later Joshi et al. (1985) inferred that Central Nepal serves as the primary source for the recurrence of rust over the Indo-Gangetic plain. However, extensive surveillance and tracking of the pathogen population is required to unveil the life cycle of *Pst* in Nepal.

The wheat yellow rust pathogens have several pathotypes reported in Nepal (Ali et al., 2018). Whether sexual reproduction of this fungi exists in Nepal is unexplored. In this regard *Berberis* plants were surveyed for yellow rust "hotspot" regions at Godabari, Phulchoki, Nagarkot, Dhulikhel, Kabre and several other places in Kabhrepalanchowk, Doti, Dadeldhura, Kathmandu, Lalitpur and Bhaktapur districts, etc. However, many (more than 25) species of *Berberis* spp., *Mahonia* spp. have been prevalent in Nepal, and these may be host to various *Puccinia* spp., as reported in Pakistan (PPD Ann. Rep. 2014/2015; Ali et al., 2015). Different *Berberis* species have been widely distributed in hills throughout the country, however, most surveyed plants had no pycnia and aecia of rust fungi in the surveyed regions during surveyed time of Feb– April (Falgun to Chaitra). Very few plant had pycnia and aecia observed in certain place like Gadabari, Phulchoki, Lalitpur. Only two *Berberis* species (*B. asiatica* and *B. aristata*) were identified from surveyed locations. Two species of *Mahonia* were also recorded at Godabari, Phulchoki, Lalitpur. Aecial cups were observed on *B. asiatica* at Botanical garden, Godabari and vicinity areas.

To know the role of sexual cycle for its contribution to the diversity of *Pst* and other rust pathogens in Nepal and its role as an alternate host, several aecia from *Berberis* plants from Godabari area were inoculated on several varieties of wheat, barley, rye, grasses and other forest tree like pine, etc. There were excellent aeciospores landed on these tested crop plants but no infection was established on these tested plants. It suggested that either these aeciospores are not the pathogen of these hosts or the experiment could not provide an adequate environment for infection. Continued studies are required for rust epidemiology and further investigation is ongoing (PPD Ann. Rep. 2014-2015).

It will also be important to see whether the pathogen on grasses and *Berberis* species is the same as on wheat crop, not only to the level of species but also to the level of virulences and pathotypes. It is possible that rust pathotypes distributed in the highlands of Nepal may spread to South Asia and other European countries. Multilateral collaboration will further confirm such possibilities, which need for a joint effort at regional level (Mahto et al., 2001).

2.3 Previous Epidemics and Resistance Break-down

In Nepal, the yellow rust was first recorded in 1964 and it remained under control when RR 21, a rust resistant wheat cultivar was popularly grown in both plain and hilly areas. In the mid of 1980s, the virulence of the rust pathogen began to change and rust epidemic was encountered due to prevalence of 7E150 race, which rendered the RR21 susceptible (Sharma et al., 1995).

Subsequently continuous researches were carried out with an objective to manage the newly emerged yellow rust pathotype and some high yielding and resistant genotypes were developed. As a result, wheat varieties Annapurna 1, Annapurna 3 and Annapurna 4 with *Yr9* gene and Annapurna 2 and Kanti with *Yr2* gene were released. Later in mid-1990s, the *Yr9* virulent race was identified in the region (Prashar et al., 2007). The wheat varieties like Annapurna 1, Annapurna 4 and Kanti resistant to the race 7E150 were rendered susceptible by yellow rust due to the appearance of a race 46S119 in Baglung, Parbat, Myagdi and other districts during 1996-1997 (Sharma, 2001; Karki et al., 1986). Currently, several yellow rust resistant wheat cultivars having *Yr27* gene located on chromosome 2BS (Singh et al., 2004) have been released for cultivation in different areas. However, the appearance of a race 71E32, rendered the *Yr27* containing CIMMYT germplasm/genotypes ineffective while resistance gene *Yr9* found effective against this pathotype (PPD Ann. Rep, 2008-2009). The year-to-year appearance of different races of yellow rust of wheat indicates that the fight against this devastating disease of wheat is far from over and warrants an attention on monitoring of pathotypes/races in order to develop genotypes with durable resistance and avoid epidemics of disease.

2.4 Wheat Cultivars Resistance Assessment

Forty-two bread wheat varieties (*T. aestivum*) and two Durum wheat (*T. durum*) have been released in different years till 2019 by Govt. of Nepal, MoALD, National Seed Board (NSB), National Wheat Research Program, NARC. Among them, 26 bread wheat varieties were released for plain and 16 were for hilly region. However, two Durum wheat have also been released for western plains during 2017 (Table 2.1).

Only 33 varieties (19 for plain and 14 for hills) are in cultivation whereas 13 varieties have been de-notified (www.sqcc.gov.np； Gairhe et al., 2018). National Seed Board (NSB) has released two Ug99 resistant varieties proposed by National Wheat Research Program (NWRP): Danphe and Tilottamaviz; NL1064 (Danphe) and NL1073 (Frankolin) for hills and plain regions respectively in the year 2015. NL1064 was resistant to major diseases like yellow rust, leaf rust and stem rust (Ug99) and also reported to be tolerant to *Helminthosporium sativum* (HLB). Similarly, Tilottama (NL1073) is resistant to yellow rust, leaf rust and stem rust (Ug99) and foliar blight *Bipolaris sorokiniana* (FB) (DhrubThapa per. Communication). Similarly, NSB has released two new wheat varieties proposed by NWRP, Swargadwary (BL3629) and Bandganga (BL3623) for hills and plain regions respectively in the year 2016 (Ann. Rep. 1975 to 2020).

Further germplasm has been developed using both exotic and local lines. A set of 982 lines was maintained in the crossing block and 58 new combinations using single, top and some three way and multiple crosses, which involved selected parents to incorporate specific desirable traits. These will lead to development of new varieties, with the major aim of rust resistance.

Table 2.1 Recommended Nepalese wheat & Durum varieties released over several years from 1960 to 2020

No.	Cultivars	Parentage/Pedigree	Origin	Release year	*Yr* gene (s)	Recommend Domain
1	Lerma -52	MENTANA/KENYA324	Mexico	1960	*Yr6+Yr7+*	Hills
2	Lerma -64	Y50/N10B//L52/3/2*LR	Mexico	1967	*Yr2+*	Hills
3	Sonora 64	YT54/N10B//2*Y54	Mexico	1967	—	Hills
4	Kalyansona	PJ "S" /GB55	Mexico	1968	*Yr(2+KS+),Yr18+*	Plains
5	Pitic 62	YT54/N10B 126.16	Mexico	1967	—	Hills
6	RR 21	II54/388/AN/3/YT54/N10B//RL64	Mexico	1970	*Yr2, YrA+*	Plain
7	NL 30	HD8325-5-5-0Y/BB	India	1975	—	Western Plain
8	HD 1982	E5557/HD845	India	1975	—	Western Plain
9	UP 262	S308/BAJIO66	India	1978	*Yr2, Yr18+*	Plains
10	Lumbini	E4871/PJ62	India	1981	*Yr2(HVII)*	Plains
11	Triveni	HD1963/HD1931	India	1982	*Yr2, Yr18+*	Plains
12	Vinayak	LC 55	India	1983	*YrGA+*	Plains
13	Sidhartha	HD2092/HD1962//E4870/K65	India	1983	—	Plains
14	Vaskar	TZPP/PL//7C	Mexico	1983	*Yr18*	Western Plains
15	Nepal 297	HD2137/HD2186//HD2160	India	1985	*Yr2+*	Plains
16	Nepal 251	WH197/HD2160//WH147	India	1988	*Yr2(HVII)*	Plains
17	Annapurna-1	KVZ "S" BUHO//KAL/BB	Mexico	1988	*Yr7,Yr9*	Hills
18	Annapurna-2	NPO/TOB "S" //8156/KAL/BB	India	1988	*Yr(2+KS+)*	Hills
19	Annapurna-3	KVZ "S" BUHO//KAL/BB	Mexico	1991	*Yr7,Yr9*	Hills
20	BL 1022	PVN "S" /ALONDRA "S"	Nepal	1991	*Yr7,Yr9*	Western Plains
21	Bhrikuti	CMT/COC75/3/PLO//FURY/ANA75	Mexico	1994	*Yr(2+KS+),Yr9,Yr18*	Plains
22	BL 1135	QTZ/TAN	Nepal	1994	*Yr9+YrSU*	Plains
23	Annapurna-4	KVZ/3/CC/INIA//CNO/ELGAU/4/ SN64	Mexico	1994	*Yr9*	Hills
24	Achyut	CPAN168/HD2204	India	1997	*Yr9*	Plains
25	Rohini	PRL "S" /TONI//CHILL "S"	Nepal	1997	*Yr18+*	Plains
26	Kanti	LIRA/FUFAN17//VEE "S"	Mexico	1997	*Yr9,Yr18*	Hills
27	Passang Lahmu	PGO/SERI	Mexico	1997	*Yr7,Yr9+*	Hills
28	BL 1473	Nepal 297/NL 352	Nepal	1999	*Yr2+Yr7+YrSU*	Plains
29	Gautam	SIDDHARTHA/NING8319/NL297	Nepal	2004	R	Plains
30	WK 1204	SW89-3064/STAR "S"	Mexico	2007	*YrA+*	Hills
31	Aditya	GS348/NL746//NL748	Nepal	2009	*YrA+*	Plains
32	NL 971	MRNG/BUC//BLO/PVN/3/PJB81	Mexico	2009	*YrA+*	Plains
33	Vijay	NL748/NL837	Nepal	2010	*Yr2+*	Plains
34	Gaura	NL872/NL868	Nepal	2012		Hills

(continuous)

No.	Cultivars	Parentage/Pedigree	Origin	Release year	*Yr* gene (s)	Recommend Domain
35	Dhaulagiri	BL 1961/NL867	Nepal	2012		Hills
36	Tilottama (NL 1073)	WAXING*2/VIVITSI/2*SERI-1B	Mexico	2015		Plains
37	Danphe (NL 1064)	KIRITATI//2*PBW65/2*SERI-1B	Mexico	2015		Hills
38	Bandganga (BL 3623)	XIA984-10YAASKUNMING/BL1868	Nepal	2016		Plains
39	Sworgadwari (BL 3629)	XIA984-10YAASKUNMING/BL1868	Nepal	2016		Plains
40	Munal (NL 1055)	WAXWING*2/KIRITATI	Mexico	2017		Hills
41	Chyakhura (NL 1164)	WHEAR/VIVITSI/3/ C80.1/3*BATAVIA///2*WBLLI	Mexico	2017		Hills
42	Khajura Durum 1	WDRAIL.I/TOSKA 26//PLATA 6/ GREEN 17/3/SORA/2*PLATA_12// SOMAT_3/4/SORA/2*PLATA 12// RASCON 3 7	Mexico	2017		Western Plains
43	Khajura Durum 2	MINIMUS_6/PLATA_16//IMMER/3/ SOOTY_9/RASCON_37/9/ USDA595/3/D67.3/RABP/CRA/4/ ALO/5/HUI/YAV 1/6/ARDENTE/7/ HUI/YAV79/8/POD9	Mexico	2017		Western Plains
44	BL 4341	—	Nepal	2018		Plains

2.5 Resistance Genes Used and the One Still Exploited/Effective

Nine *Yr* genes, i.e.,*Yr2, YrSK, YrA, Yr6, Yr7, Yr9, Yr27, YrGA*, and *YrSU* were postulated in several wheat lines of Nepal (Mahto, 1996; Mahto et al., 1999; Mahto et al., 2013). Most of resistance gene(s) were postulated in alone and some of them were in combination with two or more genes. *Yr9* was the most commonly identified resistance gene. *Yr9* is linked with *Lr26* and *Sr31* genes (Mahto, 1996; Sharma, 1997). Most of the tested lines have resistance based on one or two genes which makes most of the wheat lines vulnerable to yellow rust of wheat in the country. The genes *Yr1, Yr2, Yr6, Yr7, Yr8, Yr15, YrSU* and *YrA+* are ineffective in Nepal whereas *Yr4, Yr5, Yr9, Yr10, Yr27, YrSP* and *YrSD* are still effective against the many pathotypes of yellow rust pathogen (Mahto and Baidya, 2012a,b,c,d,e; Mahto et al., 2016).

Seedling resistance test (SRT) revealed that most of the genotype showed resistant reaction to the pathotypes of *Pst* with virulence against *Yr9*, showing avirulent infection type (IT) reaction of zero to fleck (0; to ;-) to pathotypes 13 (67S8), K (47S103) and L (70S69) whereas it had higher IT reaction to 46S119 and 79S84. The pathotypes 13 (67S8), 31 (67S64), 14A (66S64), N (46S64) and L (70S69) infected lines with *Yr2* whereas higher IT reaction was noted against pathotypes 46S119, K (47S102), 20A (70S64) and T (47S103). The Gene *Yr27* was effective to pathotypes T, I, 13, 31, K, N, 14A, 20A and 28S84. Four different resistance genes were postulated in the Nepalese wheat material viz, *YrA, Yr2, Yr9* and *Yr27* in the seedling test (Sharma, 1997). These genes were observed either singly or in combination with other gene(s). The gene *YrA* and *Yr2* were detected alone whereas *Yr27* was postulated in combination with *Yr9*. The gene *Yr9* was identified in both single and in combination to other genes. *Yr27* was postulated only in combination with *Yr9* gene. Likewise, *Yr2* gene alone was inferred in 39 genotypes (31.2%). Gene *YrA* was also detected alone in 9 genotypes (8.8%). Among them *Yr9* gene was predominantly postulated in 48 genotypes (38.4%) either singly or in combination with *Yr27* in 11 genotypes (9.6%) whereas in WK 1686, it was postulated with other resistant gene(s) (Baidya et al., 2018).

Wheat genotypes lacking seedling resistance or possessing genes for which virulence(s) were available were subjected to adult plant resistance tests in polyhouse (Ann. Rep., 2008-2016). These genotypes were classified into three categories based on the yellow rust response at seedling and adult plant stages. The first group had 21 wheat genotypes showing susceptibility at seedling stage but resistance response at adult plant stage. The genotypes were found highly resistant to both the predominant and virulent pathotypes at adult stage (Baidya et al., 2018). The second had included 16 genotypes (including Gautam), which had resistant reaction at both seedling and adult plant stages for both the pathotypes. The third group was comprised of 43 genotypes (including Achyut, Aditya, Bandganga, BL1473, BL3472, Danphe 1, Hanse, Munal 1), which were susceptible to both the pathotypes. Wheat genotypes BL 3472, Hanse, Munal-1, BL 3555, SetoGahun, WK1481, WK1712, WK1723, WK 1792 and WK1974 showed more susceptibility to pathotype 78S84 than other genotypes (Baidya et al., 2018). Similarly, Hanse, WK936, WK1544, WK1712, WK1776, WK1792, WK 1915, WK1996 and BL4040 were more susceptible to pathotype 46S119 than other lines. Among these genotypes, WK1792 and WK1996 were found 100 percent susceptible to both the pathotype (Baidya et al., 2018).

The seedling tests have also confirmed that *Yr9* was the dominant resistance gene in most of the genotypes as shown in Table 2.2, 2.3 (Karki, 1994; Mahto, 1996; Sharma, 1997; Mahto and Prashar, 1999). Most of the Nepalese wheat genotypes had crosses with CIMMYT-derived germplam/genotypes which have wide presence of *YrA, Yr9* and *Yr2* genes (Badebo et al., 1990; Danial et al., 1994). The genotypes which contained *Yr9*, based on the 1B/1R translocation, also indicated the presence of tightly linked genes *Sr31* and *Lr26* (Zeller, 1973; McIntosh et al., 1995). *YrA* gene is temporarily designated yellow rust resistance gene in a differential line Anza. The genotype RR-21 was considered to possess *YrA* and *Yr2* genes (Wellings et al., 1988; Singh et al., 1990).

The use of pathotypes to detect resistance genes has been implied for the Nepalese germplasm. Such studies showed that *Yr2* and *Yr9* are still effective genes against common pathotypes of *Pst* in Nepal. However, another report showed that genes, *Yr1, Yr2, Yr6, Yr7, Yr8, Yr15, YrSu* and *YrA+* are ineffective in Nepal whereas *Yr4, Yr5, Yr9, Yr10, Yr27, YrSP, YrSD* are still effective (Sharma et al., 1995; PPD, 2009). Similarly, the *Yr4* gene was suggested to be ineffective in India (Kumar et al., 1993) and Pakistan (Lourwers et al., 1992). The response of resistance genes in the field, however, vary with the evolution of new pathotypes in different time intervals at different locations.

The gene *Yr27* was found ineffective against 78S84 and 46S119 pathotypes because the genotypes which possessed *Yr27* were susceptible to both the pathotypes at adult stage. Likewise, *YrA* gene was also ineffective to pathotype 46S119, however, the genotypes BL2879, WK1204, Chonte which contained *YrA* gene were resistant to both the pathotypes and it could be due to additive effect of other genes presence in genotypes. Fifteen genotypes (BL1473, BL3063, BL3468, BL3978, BL4095, BL4118, BL4154, NL1073, PasangLahmu, WK1628, WK1710, WK1909, WK1970, WK1998 and Chonte 1) also showed the presence of slow rusting gene *Lr34* on the basis of phenotypic characteristics. Phenotypic marker can be observed in the genotypes which possess *Yr18/Lr34* genes generally produce broken yellows infected by *Pst* at adult plant stage and progressive leaf tip necrosis (Nayar et al., 2001). The gene *Yr18* is commonly known to confer slow rusting response. So, these genotypes could be utilized in the breeding program for developing durable resistance. Most of the genotypes have resistance based on one or two genes which makes vulnerable to yellow rust of wheat in the country. The pattern of evolution of the pathotypes of *Pst* in the country showed wide variation and a clear increase in the number of virulence factors. Due to the evolution of new virulences, the erosion of resistance gene(s) to the pathotypes was rapid and the formation of highly complex pathotypes was observed. The evolution of new virulence pathotype has been closely associated with *Yr*-resistance gene development in new cultivars. The combination of seedling resistance and adult plant resistance has potential to keep the initial population of the year round epidemic pathotypes in low level. The selection of adult plant resistance in hypersensitive and non-hypersensitive genotypes could impart lasting resistance. Varietal diversification and identification of different resistance genes for different environments are needed for better management of yellow rust.

Table 2.2 Postulation of *Yr*-gene(s) in selected wheat lines/varieties from Nepal

Gene(s)	Lines	Test lines/varieties
Yr(2+KS+)	2	Bhrikuti, Annapurnna-2
Yr2 (SKA)+	5	BL 1496, NL 773, NL 715, UP 262, Triveni
Yr9	24	NL714, NL727, NL728, NL747, NL750, NL753, NL764, NL769, NL771, NL776, NL781, NL783, NL784, NL785, NL805, WK810, WK820, BL1022, BL1655, BL1658, Annapurna-1, Annapurna-3, Annapurna-4
R to All Pathotypes	6	NL731, NL748, NL758, NL794, NL800

Table 2.3 Comparative Postulation of *Yr* genes in Nepalese wheat lines at DWR, Shimla, H.P. India during 1994—1996

No.	Varieties	1994	1996
1	Nepal 297	*Yr2(HVII)*	—
2	Nepal 251	*Yr2(HVII)*	—
3	Triveni	*Yr2(HVII)*	*Yr2(SKA)*
4	Siddhartha	—	—
5	Bhrikuti	*Yr9*	*Yr9*
6	Annapurna-1	*Yr9*	*Yr9*
7	Annapurna-2	—	—
8	Annapurana-3	—	—
9	Annapurna-4	*Yr9*	*Yr9*
10	BL1135	—	—
11	RR21	—	—
12	UP262	*Yr2 (HVII)*	*Yr2(SKA)*
13	Vinayak	—	—
14	Vaskar	—	—
15	BL1022	—	*Yr9*
16	Lumbini	*Yr2 (HVII)*	Not tested
17	BL1473	—	—
18	BL1496	*Yr2 (HVII)*	*Yr2(SKA)*
19	BL1530	*Yr9*	—
20	WK810	*Yr9*	*Yr9*
21	WK820	—	*Yr9*

2.6 Virulence Variation and Resistance Breakdown

The yellow rust pathogen is highly variable in Himalayan region, which has been suggested as the center of origin of the pathogen (Ali et al., 2014a). Like other Himalayan populations, many pathotypes are known to occur in Nepal (Ali et al., 2017; Ali et al., 2018). The evolution of these pathotypes renders the existing resistant variety susceptible. A rust infected field of a susceptible variety may have more than one pathotypes in these regions with high diversity (Ali et al., 2014b).

In Nepal, more than 29 pathotypes of *Pst* have been recorded till 2020. During 70's and 80's, the pathotyeps 3E0, 7E0, 7E2, 7E6, 7E10, 7E130, 7E134, 7E130, 6E0, 4E16, 2E0, 68E16, 7E158, 15E150, 15E158, 70E0, 66E18, 134E150, and 7E150 were observed from different part of the country (Table 2.4 and 2.5). Pathotypes L (70S69), P (46S103), K (47S102), 14A (66S64), 20A (70S64) and T (47S103) which were virulent to *Yr2*, were isolated in 90's (Karki 1980; Karki 1989). Three pathotypes, L, P and T were also virulent to *Yr1, Yr2, Yr3a, Yr4b, Yr6, Yr7, Yr8, YrKS, YrSD* and *YrSKA*. Pathotype 47S103 (47E148) was detected during 1995 from central region of Nepal. Likewise, pathotype 46S119 was detected in 1997 from western region which possessed virulence for *Yr9* and it was considered to have arisen from a single-step mutation in pathotype 46S103. In 2000, pathotype 78S84 rendered *Yr9* and *Yr27* susceptible. Four pathotypes 46S103 (46E151), 46S119 (46E151+*Yr9*), 78S84 (78E16) and (71E32) have been predominating in the yellow rust population during last couple of years, while others were found in low frequency.

New virulence in the existing pathotypes may arise through either mutation, somatic hybridization or recombination. The pattern of evolution of the pathotypes of *Pst* in Nepal showed wide variation and a clear increase in the number of virulence factors. Genetic resistance is the most economical and eco-friendly method of reducing yield losses due to wheat rusts. Ug99 resistant wheat variety Vijay had been also released for cultivation during 2010. Afterwards several other Ug99 resistance varieties like Danphe etc. have been released so far. Nine *Yr* pattern (*Yr2*, *YrKSA, YrA, Yr6, Yr7, Yr9, Yr27, YrGA,* and *YrSU*) have been postulated in Nepalese wheat lines. Most of them were postulated alone and few were in combination with two or more genes which make most of the wheat lines vulnerable to yellow rust pathotypes. Considering the overall pathotyping studies, *Yr4, Yr5, Yr9, Yr10, Yr27, YrSP* and *YrSD* have been found effective over the many pathotypes of *Pst* in Nepal. The high prevalence of virulence to *Yr9* could be due to the fact that it is the most commonly identified resistance gene and linked to *Lr26* and *Sr31* genes (Mahto and Baidya, 2012a, b, c ,d; Mahto et al., 2016).

Population genetic studies on the Nepali population have revealed a high diversity with some signature of recombination in the Nepali population (Ali et al., 2018; Khan et al., 2019). The Nepali population was completely distinct from Pakistani population (Khan et al., 2019), but still highly diverse. The Nepali isolates have been shown to have high sexual reproduction capacity (Ali et al., 2010). The possibility of sexual reproduction on alternate host, i.e., *Berberis* spp. in the country could not be ignored due to prevalence of several *Berberis* spp. Recent research based on weather data found that Nepal has mid-risk of *Berberis* infection by *Pst*, the infection score was 0.67 (Sinha and Chen, 2021). It was therefore necessary to confirm the role of alternate host for evolving new pathotypes of yellow rust pathogens in the country (Mahto and Baidya, 2012a,b,c,d; Mahto et al., 2016).

The change in virulence spectrum of rust pathogens may be caused by higher selection pressure *i.e* change in wheat varieties or genotypes grown in the country. The resistance of a line/ genotype with single effective resistance gene can easily overcome by one step mutation in the pathogen.

Table 2.4 Pathotypes prevalent in the Nepalese yellow rust population and the varieties which are susceptible to these as detected during 1970-2012

No.	Year	Pathotype	Remark
1	1970 - early 1980	2E0, 3E0, 6E0, 4E16	Premitive Pathotype
2	1985 onward	7E150, 15E150, 15E158	Widely cultivated variety Sonalika become highly susceptible
3	1997 onward	46S119(*Yr9* virulence)	Genotypes: Annapurna-1, Annapurna-3 and Annapurna-4, BL 1022, having *Yr9* Susceptible
4	2005 onward	71E32 Major Yellow rust epidemic (field observation)	—
5	Current	78S84	Has not caused any epidemic DWR (Shimla, 2011)

Table 2.5 Pathotypes prevalent in the Nepalese yellow rust population monitored since 1970s

No.	Yellow rust pathotypes	Year of origin
1	3E0, 7E0, 7E2, 7E6, 7E10, 7E130, 7E134, 7E130, 6E0, 4E16, 2E0, and 68E16	1970's–early 1980's
2	7E158, 15E150, 15E158, 70E0, 66E18, 134E150 and 7E150	Late 1980's
3	L (70S69), P (46S103), K (47S102), 14A (66S64), 20A (70S64) and T (47S103)	Early 1990's
4	47S103 and 46S119	Late 1990's
5	71E32 and 78S84	2000's
6	46S103 (46E151), 46S119 (46E151 + *Yr9*), 78S84 (78E16) and 71E32)	2001–2020

2.7 Molecular Studies on Resistance Genes

Very limited work has been done in molecular aspect of rust resistance in wheat lines in Nepal. The study based on SSR markers showed that *Lr34* gene was present in 32% of tested Nepalese lines, the frequency of *Lr19* was 1.9%, *Sr2* was 28%, *Sr22* was 38% and *Sr36* was 7% (Baidya et al., 2013; Baidya et al., 2014; Baidya et al., 2019). None of the wheat lines had *Yr36* gene, when assessed with SSR UHW89 markers. A more comprehensive work is thus needed to assess the presence of various rust resistance genes in the Nepalese wheat germplasm.

2.8 Future Avenues to Work on

Identification and transfer of new sources of race-specific and race non-specific resistance from various wheat relatives would enhance the diversity of resistance. Although new yellow rust resistant varieties that yield more than current popular varieties should be released and promoted, major efforts are required to replace current yellow rust susceptible varieties with varieties that have diverse resistance sources and mitigate the yellow rust threat. In coming days, the efforts of developing and releasing the wheat varieties to cope up with the new challenges will be continued. The disease incidence changes according to the climatic condition and management practices. New virulent pathotypes of different diseases are emerging that posse new challenges to overcome through new breeding strategies by identifying resistant parents, crossing and selection to develop new resistant varieties. Moreover, time has come to click on biotechnology and molecular breeding approach to understand the pathogen population biology and host resistance in Nepalese germplasm. Research should focus on the alternate sources of resistance against wheat rusts in Nepal. Molecular markers assisted selection is an important tool for rapid pyramiding of *Lr, Yr,* and *Sr* genes. Isolation, characterization and mapping of rust resistance genes are essential to overcome this disease through effective and judicious deployment of resistance genes.

References

Ali S, Gladieux P, Leconte M, et al., 2014. Origin, Migration Routes and Worldwide Population Genetic Structure of the Wheat Yellow Rust Pathogen *Puccinia striiformis* f. sp. *tritici*. *PLoS Pathogen*, 10 (1): e1003903. DOI:10.1371/journal.ppat.1003903.

Ali S, Hovmoller MS, Swati ZA, et al., 2015. Diversity of Puccinia spp. on Wheat, Grasses and *Berberis* spp. in the Himalayan Center of Diversity of *Puccinia striiformis* f. sp. *tritici*. *BGRI Workshop Poster Abstract*.

Ali S, Leconte M, Walker AS, et al., 2010. Reduction in the Sex Ability of Worldwide Clonal Populations of *Puccinia striiformis* f.sp. *tritici*. *Fungal Genetics and Biology*, 47:828-838.

Ali S, Rodriguez-Algaba J, Thach T, et al., 2017. Yellow Rust Epidemics Worldwide Were Caused by Pathogen Races From Divergent Genetic Lineages. *Front. Plant Sci*, 8:1058.

Ali S, Sharma S, Leconte M, et al., 2018. Low Pathotype Diversity in a Recombinant *Puccinia Striiformis* Population Through Convergent Selection at the Eastern Part of Himalayan Centre of Diversity (Nepal). *Plant Pathology*, 67: 810-820. DOI:10.1111/ppa.12796.

Annual Report. International Maize and Wheat Improvement Centre, 2013. FAO. 2016. FAOSTAT. http://

faostat3.fao.org

Annual Reports 1975-2020. National Wheat Research Program, Nepal. Bhairahawa, Badebbo A, Stubbs RW, et al., 1990. Identification of Resistance Genes to *Puccinia striiformis* in Seedlings of Ethiopian and CIMMYT Bread Wheat Varieties and Lines. *Neth. J. Pl. Path.*, 96:199-210.

Anononymous, 1997. Twenty-five Years of Wheat Research in Nepal (1972-1997). Nepal Agricultural Research Council, NWRP, Nepal.

Baidya S, Bhardwaj SC, Shrestha SM, et al., 2018. Characterization of Yellow Rust (*Puccinia striiformis*) Resistance and Genetic Diversity in Nepalese Wheat Genotypes. *J. Pl. Prot. Soc,* 15:175-183.

Daniel DL, Stubbs RW and Parlevliet JE, 1994. Evolution of Virulence Patterns in Yellow Rust Races and its Implication for Breeding for Resistance in Wheat in Kenya. *Euphytica*, 80:165-170.

Eriksson, Jakob, 1984. Ueber die Specialisieruns des Getreideschwarzrostes in Jhyeden Land in Anderen Lindem. Gentbl. Bakt. (II), 9: 590-607, 654-658.

FAO, 2014. Statistical Yearbook. World Food and Agriculture. Food and Agriculture Organization of the United Nations. Rome.

Gairhe S, Shrestha H, Timsina K, 2018. Dynamics of Major Cereals Productivity in Nepal. *Journal of Nepal Agricultural Research Council,* 4(1): 60-71. https://doi.org/10.3126/jnarc.v4i1.19691 http://www.cimmyt.org/cimmyt-s-2013-annual-report-now-available-online/

International Maize and Wheat Improvement Center, CIMMYT Annual Report, 1997 to 2000, 2013. Mexico, D.F. CIMMYT.

Jin Y, Szabo LJ, Carson M, 2010. Century-old Mystery of *Puccinia striiformis* Life History Solved with the Identification of *Berberis* as an Alternate Host. *Phytopathology,* 100:432-435.

Joshi LM, Srivastava KD, Singh DV, 1985. Monitoring of Wheat Rusts in the Indian Sub-continent. *Proc. Indian Aca. Sci. (Plant Sci.),* 94: 387-407.

Karki CB, 1980. Report on Evaluation on Nepalese Wheat and Barley Varieties in the Seedling Stage on their Resistance to Yellow Rust. Research Institute for Plant Protection (IPO), Wageningen, the Netherlands.

Karki CB, 1989. Wheat Yellow Rust Epidemics and Monitoring Virulences of its Pathogen, *Puccinia striiformis* f. sp. *tritici* in Nepal. The paper presented at the Regional Seminar on Microbial Research, Dec.1-5, 1989, Royal Nepal Academy of Science and Technology (RONAST), Kathmandu, Nepal.

Karki CB, 1994. Genetics of Rust Resistance of Some Nepalese Wheat and Barley Cultivars. A Research Study Carried out at Directorate of Wheat Research (DWR), Regional Station, Flowerdale, Shimla-171002, H.P. India.

Karki CB, Sharma S, 1990. Wheat Disease Report 1989/1990. In :The Proceeding of the Thirteenth National Winter Crops Seminar (Wheat Report), Sept.10-14, 1990, National Wheat Research Programme, Bhairahwa, Nepal.

Karki CB, Sharma S, Dangol RHS, 1986. Wheat Yellow Rust Situation in Nepal. Paper presented at the First

Wheat Working Group Meeting, Siddhartha Nagar, Nepal.

Khan MR, Rahman ZU, Nazir SN, et al., 2019. Genetic Divergence and Diversity in Himalayan *Puccinia Striiformis* Populations from Bhutan, Nepal and Pakistan. *Phytopathology*, 109:1793-1800.

Kumar J, Nayar SK, Prashar M, et al., 1993. Virulence Survey of *Puccinia Striifromis* in India during 1990-1992. *Cereal Rust and Powdery Mildew Bulletins*, 21:17-24.

Louwers JM, Van Silfhout CH, Stubbs RW, 1992. Race Analysis of Yellow Rust in Developming Countries, Report 1990-1992, IPO-DLO report: 11-23.

Mahto BN, 2014. Genetics of Rust Resistance of Selected Wheat Lines and Mapping of Pathotypes Flora of Leaf Rust in Nepal. A research study carried out during July 4-Sept.15, 1996 at DWR, Regional Station, Flowerdale, Shimla-171002, H.P., India.

Mahto BN, Baidya S, 2012c. Status of Wheat Rust Diseases and Their Management in Nepal. Paper presented at South Asia Regional and National Workshop on Contigency Planning for Management of wheat rust diseases organized by FAO/MoAC/NARC/DoA from 18-21 Dec., Kathmandu, Nepal.

Mahto BN, Baidya S, 2012d. Status of Wheat Rust Diseases and Their Management in Nepal. Paper presented at South Asia Regional and National Workshop on Contigency Planning for Management of Wheat Rust Diseases organized by FAO/MoAD, NARC/DoA from 18-21 Dec, Kathmandu, Nepal.

Mahto BN, Baidya S, Sharma S, et al., 2012. Current Status, Challenges, Capabilities, Strength and Institutional Arrangements of Wheat Rust Diseases and Their Management in Nepal. Paper presented at South Asia Regional and National Workshop on Contigency Planning for Management of Wheat Rust Diseases organized by FAO/MoAD, NARC/DoA from 18-21 Dec. 2012, kathmandu, Nepal.

Mahto BN, Baidya S, 2012a. Genetic Basis of Yellow, Leaf and Stem Rust Resistance in Nepalese Wheat. Poster presented at Regional and national contingency planning for wheat rust management in Nepal. Organized by FAO/MoAD, NARC/DoA from 18-21 Dec., Kathmandu, Nepal.

Mahto BN, Baidya S, 2012b. Virulence Spectrum of *Puccinia triticina* and *P. striformis* in Nepal. Poster presented at Regional and national contingency planning for wheat rust management in Nepal. Organized by FAO/MoAD, NARC/DoA from 18-21 Dec. Kathmandu, Nepal.

Mahto BN, Nayar SK, Nagarajan S, 2001. Postulation of *Lr* Genes in the Bread Wheat Material of Nepal Using Indian Pathotypes. *Indian Phytopath*, 54(3): 319-322.

Mahto BN, Prashar M, 1999. Genetics of Yellow Rust Resistance in Nepalese Wheat. *Cereal and Powdery Mildews Bulletin*, 26:27-34.

Mahto BN, Shrestha R, Baidya S, 2014. Monitoring Survey of Wheat and Barley Diseases in the Context of Climate Change in Nepal. In Proceedings of the 29th National Winter Crops Workshop held on 11-12 June, 2014 at Regional Agriculture Research Station, Nepal Agricultural Research Council (NARC), Lumle, Kaski, Nepal.

McIntosh RA, Wellings CR, Park RF, 1995. Wheat Rusts: an Atlas of Resistance Genes. Plant Breeding Institute, the University of Sydney, CSIRO publication, Victoria, Australia 200.

Mehta KC, 1940. Further Studies on Cereal Rusts in India. Part1. *Indian Coun. Agri. Res. India. Sci. Monogr,* (14):1-65.

MoAD, 2015. Statistical Information on Nepalese Agriculture. Ministry of Agriculture Development, Singh Durbar, Kathmandu, Nepal.

MoF, 2014. Economic Survey. Ministry of Finance, Singh Durbar, Kathmandu, Nepal.

Morris ML, Dubin HJ, Pokhrel T, 1994. Returns to Wheat Breeding Research in Nepal. *Agricultural economics*, 10:269-282.

Nayar SK, Nagarajan S, Prashar M, et al., 2001. Revised Catalogue of Genes that Accord Resistance to *Puccinia* Species in Wheat. Directorate of Wheat Research, Regeional Station, Flowerdale, Shimla -171002 (India). *Research Bulletin*, 3:48.

Pokharel TP, Bhatta MR, 2001. Three Decades of NARC-CIMMYT Partnership in Maize and Wheat Research and Developmenyt (1970-2000). Nepal Agricultural Research Council (NARC) and International Maize and Wheat Improvement Center (CIMMYT), Nepal February 12, 2001.

PPD, 2008. Annual Report 2006/2007, Plant Pathology Division, Nepal Agricultural Research Council (NARC), Khumaltar, Lalitpur, Nepal.

PPD, 2014. Annual Report 2013/2014. Plant Pathology Division, NARC, Khumaltar, Lalitpur, Nepal.

Prashar M, Bhardwaj SC, Jain SK, et al., 2007. Pathotypic evolution in *Puccinia striiformis* in India during 1995-2004. *Australian J. Agr. Res,* 58:602-604.

Scliroeter J, Sntwicklxinssgesciiichte ©iniger Roatpilze. Beitr, Biol. Pflanzen. 1879;3: 51-93.

Sharma S, 1997. Virulence Monitoring and Detection of Leaf and Yellow Rust Resistance Genes in Nepalese Wheat Varieties. A Research Study carried out during July 4 to August 11. Directorate of Wheat Research, Regional Station, Flowerdale, Simla, India.

Sharma S, 2001. Yellow Rust of Wheat in Western Hills of Nepal. Advances in Agricultural Research in Nepal. Proceeding of the first SAS/N Convention. Society of Agricultural Scientist: 170-175. Indian Agricultural Research Institute (IARI), Regional Station, Flowerdale, Shimla-171002. 12p.

Sharma S, Ghimire SR, Pradhanang PN, 1995. Identification of Yellow Rust Races of Wheat in the Western Hills of Nepal. Proceedings of Wheat Research Report, NWRP, Bhairahwa, Nepal: 349-354.

Sharma S, Louwers JM, Karki CB, et al., 1995. Postulation of Resistance Genes to Yellow Rust in Wild Emmer Derivatives and Advanced Wheat Lines from Nepal. *Euphytica*, 81:271-277.

Singh H, Johnson R, Sethi D, 1990. Genes for Race-specific Resistance to Yellow Rust (*Puccinia striiformis*) in Indian Wheat Cultivars. *Plant Pathology*, 39:424-433.

Singh RP, Duveiller E, Huerta-Espino E, 2004. Virulence to Yellow Rust Resistance Gene *Yr27*: A New Threat to Stable Wheat Production in Asia. Regional Yellow Rust Conference, Islambad, Pakistan, March: 22-26.

Sinha P, Chen X, 2021. Potential Infection Risks of the Wheat Yellow Rust and Stem Rust Pathogens on Barberry in Asia and Southeastern Europe. *Plants* (Basel), 10(5):957. DOI: 10.3390/plants10050957.

Wellings CR, McIntosh RA, Hussain M, 1988. A New Source of Resistance to *Puccinia striiformis* f. sp. *tritici* in Spring Wheat (*Triticum aestivum*). *Plant Breeding*, 100:88-96.

Zeller FJ, 1973. Wheat–rye Chromosome Substitutions and Translocations. *Proceeding of 4th International Wheat Genetics Symposium, Columbia*: 209-221.

Chapter 3

Wheat Yellow Rust Status Across Pakistan – a Part of the Pathogen Center of Diversity

Sajid Ali[1]*, Zahoor A. Swati[2], Muhammad Rameez Khan[2], Aamir Iqbal[2], Zia-ur-Rehman[2], Muhammad Awais[3], Ghulam Ullah[4], Ihtisham Khokhar[4], Muhammad Imtiaz[4], Muhammad Fayyaz[5]

[1] Department of Agriculture, Hazara University, Mansehra, Pakistan

[2] Institute of Biotechnology & Genetic Engineering, The University of Agriculture, Peshawar, Pakistan

[3] Northwest Agriculture and Forestry University, Yangling, China

[4] International Maize and Wheat Improvement Center (CIMMYT), Islamabad, Pakistan

[5] National Agriculture Research Station, Islamabad, Pakistan

* Corresponding author: bioscientist122@yahoo.com

Abstract: Wheat yellow rust is one of the important diseases of wheat worldwide, with presence of huge diversity in the Himalayan and near-Himalayan region, including Pakistan. The chapter attempts to provide a comprehensive summary of various research work done on yellow rust across Pakistan. Most of the earlier research work in Pakistan was mainly focused on field testing of candidate lines and released varieties across multiple locations along with trap nurseries to assess variability in the pathogen population based on infection on the differential lines with certain known genes. In the past two decades, intensive research work has been done using molecular markers-based screening of the host genotypes, both breeding lines and released varieties. Similarly, a strong research group has been established to track the pathogen population structure, using extensive surveillance and sequencing-based genotyping. The chapter summarizes the knowledge of the disease epidemics, pathogen population structure, resistance gene in host germplasm and integrated management of rust in Pakistan. The information should be useful for future resistance gene deployment and disease management, not only in Pakistan but also at regional and global scale.

Keywords: Wheat Yellow Rust; *Puccinia striiformis* f. sp. *tritici* (*Pst*); Resistance; Pathogenicity

3.1 Wheat Cultivation in Pakistan

Wheat crop pathogens, particularly cereal rusts possess severe threat to world food security and sustainable agricultural systems (Dean et al., 2012; Beddow et al., 2015). Severe yield losses by rust disease projecting to millions of tons have been reported in a single season in different parts of the world (Line, 2002; Chen, 2014). Wheat yellow rust is one of the important diseases of wheat worldwide, with presence of huge diversity in the Himalayan and near-Himalayan region, including Pakistan (Ali et al., 2014a). This highly diverse rust pathogen populations regularly infect wheat crop in Pakistan, where more than 8 million ha land is under wheat cultivation under diverse cropping systems.

The wheat cropping system in Pakistan has some specific features. It is mostly a winter crop grown under diverse ecological conditions ranging from small contour plots in Himalayan valleys to the large-scale fields of Punjab and Sindh near the Indian Ocean (Fig. 3.1). Interestingly, spring wheat is grown under winter crop season (October/November to April/May) in most of these wheat growing regions, except in the high-altitude regions of Gilgit-Baltistan and Chitral, where spring wheat is grown as spring crop (March/April to July/August). Winter wheat is rarely utilized by these farmers, despite their potential, both as grain and dual-purpose crop (Ali et al., 2009). Commercial exploitation is made in Punjab and Sindh, while in Himalayas, subsistence farming is the routine, where the crop is more diverse with several weeds, including indigenous grasses. The wheat cropping system in the Himalayan regions is often on small contour fields, where occasionally these are "protected" by various shrubs, including *Berberis* spp., which is the alternate host of wheat stem and yellow rust pathogens (Jin et al., 2010; Ali et al., 2014b; Sinha and Chen, 2021). In small land holding, one of the objectives of the crop is to provide fodder for the livestock and thus the female farmers regularly remove weeds as livestock feed. In such cropping system, little emphasis is given in the start to homogenous stand. Interestingly in some of the fields, a mixture of different landraces along with other grasses could be observed, where the crop is grown only as a fodder.

Wheat crop is cultivated in many cropping systems including wheat-maize, wheat-rice, wheat-cotton, wheat-vegetables etc. However, wheat production is impacted by many abiotic stresses including terminal heat stress during some years and high rainfall during the grain filling and maturity, which impact the decision regarding wheat sowing, which has direct impact on yield as well the onset of rust diseases.

Fig. 3.1 Wheat cultivation under diverse agro-ecological zones across Pakistan, ranging from high altitude valleys in Himalayas to big scale fields near the Indian Ocean

Despite the cultivation of wheat on a large acreage (8 million ha), the average yield per ha is much lower than other wheat producing nations. Biotic stresses, particularly wheat rusts are the major wheat constraints (Hovmøller et al., 2010; Ali et al., 2014a). To combat the disease, various disease control measures are available, but host resistance is the only major control measure exploited so far in Pakistan. This, however, has several limitations, considering the high diversity in the pathogen population (Ali et al., 2014b,c; Khan et al., 2019). The work on pathogen population has revealed that Pakistan is in the pathogen centre of diversity and potential centre of origin, with potential role of its alternate host, i.e., *Berberis* spp. (Jin et al., 2010; Khan et al., 2020). Despite this high diversity of the pathogen, almost all of the varieties are based on the germplasm received from CIMMYT, which make huge contribution to the wheat production in Pakistan, though with the limitation of narrow genetic background. Thus, wheat genetic improvement should involve exploitation of diverse genetic resources with a more justified utilization of resistance varieties at the regional and national level. Novel sources of germplasm should be explored, e.g., Australian, North African and Chinese germplasm as well as landraces from the Middle East.

3.2 Disease Onset and Prevalence of Rust Epidemics

Wheat crop has been reported with several diseases in Pakistan, e.g., Rusts, Smut, Bunts,

Powdery mildew, Septoria leaf blotch and Fusarium Head blight etc., however, rust diseases are the most important diseases. At least one of these three rusts has been reported in all wheat growing regions of the country. Among these three rust diseases, wheat stem rust, caused by *Puccinia graminis* f. sp. *tritici* is rare in Pakistan and found only in the southern part of Pakistan in a very low severity (Fig. 3.2). Wheat leaf rust, caused by *P. triticina* and wheat yellow rust remains the major threats to wheat production in Pakistan. Wheat leaf rust is more prevalent in the southern half of the country in the Sindh province, Southern Punjab and often in central and northern Punjab, depending on the prevalent climate. Yellow rust has been a major problem in northern half of the country with cold temperature and high humidity from January to April, depending on the year. The incidence and distribution of the three rust varies across years, depending upon the climatic conditions and susceptibility of the varieties deployed at the field level. In some years yellow rust was reported in high severity in the southern part of the country (Sindh), similarly in other years leaf rust becomes important in the north, where the disease is normally not considered as a risk. In addition to rusts, powdery mildew is also one of the important, but least studied, disease in the northern part of the country as well as in the central part during the years with extended cold and humid climatic conditions.

Wheat yellow rust has been reported in all wheat growing regions of the country, though the disease infestation varies over years (Fig. 3.2). The Himalaya region of Pakistan has been shown to be the pathogen centre of diversity, where the pathogen could complete its sexual cycle on the *Berberis* spp. These could serve as the regions where the pathogen could over-summer and over-winter for initiating the subsequent year epidemics. The pathogen could then spread from the Himalayan region to other parts of the country. Although a comprehensive study is required to confirm the exact within- and across-season migration pattern, indirect evidence on pathogen population structure has revealed lack of any clear geographical population subdivision with clear evidence of migration (Ali et al., 2014b; Khan et al., 2019; Khan et al., 2020).

The wheat rusts are disseminated from one part to other via different ways but wind plays an important role for frequent long-distance dispersal. The fungus is capable of long-distance migration, with well-documented cases of recurrent reestablishment of pathogen populations in areas where there are no host plants during summer/winter to allow the pathogen survival, as for the main wheat-growing provinces of north-eastern China. Such spread can be due to successive jumps from field to field by this polycyclic disease throughout the season (Brown, 2002; Hovmøller et al., 2016). Human traveling may also act as media for long distance dispersal of pathogen as exemplified in several cases like the spread of yellow rust into North American and Australia, or the spread of different pathogen lineages (Ali et al., 2014a; Ali et al., 2017a). To explore the relative importance of various factors for pathogen dispersal and the detailed

Fig. 3.2 Yellow rust infection on wheat crop observed under diverse agro-ecological zones across Pakistan

pattern of migrations across Pakistan, further studies would be required with more intensive and extensive surveillance and more informative molecular markers along with indirect observations of the disease on wheat, grasses and *Berberis* spp.

3.3 Rust Epidemics in the Past and Associated Losses

Both leaf and yellow rust has been shown to impart severe economic losses to wheat production in Pakistan, stem rust remained negligible while considering economic losses. Wheat yellow rust has been shown to result in economic losses to national wheat yield over the past several decades.

Yellow rust of wheat is among the most important foliar diseases of wheat and 70% of the acreage under wheat is conducive to it in Pakistan. Its regular outbreaks in the country have caused severe losses to national economy. National losses of 1978 and 1995 alone were more than US$116 million (Roelf and Bushnell, 1985; Saari et al., 1995). A damage of Rs. 1.5 billion was estimated to be imparted on the economy if one percent loss occurs in wheat yield (Kissana et al., 2003). The agro-climatic conditions of Pakistan, in general and that of the Himalayan region, in specific, are favourable for the pathogen of this disease. These epidemics were resulted from the onset of favorable season and from the fact that certain widely grown set of closely related varieties became susceptible.

Rust resistance in improved wheat cultivars in the past was solely based on major genes that were often overcome by virulent pathogenic mutant strains within five to seven years (CIMMYT, 1992; de Vallavieille-Pope et al., 2012). This was the case in northern part of Pakistan in mid 1990s, when *Yr9* gene based two popular wheat cultivars, namely, Pirsabak-85 and Pak-81 became susceptible and resulted in yield losses up to 40% (Saari et al., 1995). Later in early 1990s, the newly developed varieties like Inqilab-91, based on yellow rust resistance gene *Yr27*, occupied 60% of wheat production area in 1990s in Pakistan, while over 70% in Khyber Pakhtunkhwa. In the neighbouring countries like India (PBW343) and Nepal, wheat varieties protected by the same resistance gene *Yr27*, were widely deployed in the region. This large-scale exploitation of *Yr27* based varieties resulted in selection of virulence against *Yr27* and resistance of Inqilab-91 was compromised resulting in the widespread epidemics in early 2000s (Kissana et al., 2003).

3.4 Resistance Genes Incorporated into the Pakistani Wheat Germplasm

The use of genetic resistance remained the only practical yellow rust control measure at the country level in Pakistan. The development of rust resistant varieties is one of the major aims of wheat breeding in Pakistan. The rust infestation score at multiple locations across the country is one of the criteria for approval of any newly developed line under the variety approval procedures. Thus, breeding for rust resistance is a major part of wheat genetic improvement.

Development of wheat varieties relied mostly upon the lines received from CIMMYT, with many varieties developed based on one or few major resistance gene(s) (CIMMYT, 1992). Manzoor Bajwa is considered to be the pioneer for introducing the semi-dwarf wheat lines into South Asia and thus contributing to the "Green Revolution" . Since then, wheat breeding programmed in Pakistan is strongly dependent on CIMMYT lines. Most of these became susceptible in a short period of time, from one to ten years, as in other parts of the world (de Vallavieille-Pope et al., 2012; Ali et al., 2014c). This was experienced in *Yr9* based cultivars, during 1990s (Saari et al., 1995) and *Yr27* based cultivars (Inqilab-91, in Pakistan and PBW343, in India) (Kissana et al., 2003). The most recent example is of Galaxy-2013, which was reported to be susceptible just the year after release.

An effort to identify the distribution of some of known resistance genes in recently released Pakistani wheat varieties/lines revealed differential prevalence of various rust resistance genes in different wheat material analyzed. Considering these studies, it could be noticed that *Yr17* was the most prevalent yellow rust resistant gene in the current wheat varieties in Pakistan (Table 3.1). Yellow rust resistance gene *Yr18* and *Yr27* was also present in a good proportion of lines. The studies targeting exotic wheat material, which is under consideration for varietal development, revealed a different distribution of resistance genes with some of genes like *Yr5* and *Yr29* to be more prevalent. Such lines could be utilized to diversify the resistance background of wheat varieties in Pakistan.

The breakdown of rust resistance in recently released varieties has been reported to be quite fast during the past decade. One possible reason for this rapid loss of resistance could be due to wide deployment of few varieties, which has been encouraged during the recent moves by the government to provide improved and certified seed to the farmers. The best preforming varieties, but only few in number, are selected, multiplied and disseminated to the farming community. Thus, the variability present at the farmer field in the past, has been substantially reduced, including the loss of landraces, fields with mixed lines and cultivation of different varieties. In addition, most of the seed multiplication farms are in the southern part of the country and thus not the hotspot for wheat yellow rust, where even the susceptible varieties show no infection in the multiplication plots. However, the seed is distributed in the north as well, where the varieties show strong susceptibility. This encouraged a high disease pressure on these few selected varieties, which not only impart economic losses, but also provide opportunity for the pathogen to increase its inoculum. As multiple lineages could infect the same variety in this diverse and recombinant area, it could result in a rapid break-down of host resistance in the recently released varieties.

Diversification of resistance background and overall diversity in wheat varieties of Pakistan is highly important. There is a dire need to exploit novel sources of resistance along with an overall diversification of wheat germplasm using wheat lines from sources other than CIMMYT material. Additionally, the information regarding resistance genes and corresponding virulence in the neighboring countries is also important, considering the long-distance migration capacity of the pathogen.

Another important strategy is the replacement of old varieties with the improved resistant varieties should be encouraged on regular basis, instead of few varieties which outperformed in first few years and then continued over decades, despite their susceptibility. Most of the seed industries are in the south, where rust is not a major problem, and thus these susceptible varieties multiplication trials also get approved. These, however, contribute to epidemic development in other parts of the country and thus should be discouraged from multiplication. Cultivation of few lines, despite the availability of many improved varieties, as observed for the Sindh province, should also be discouraged. It will require involvement of all stakeholders from breeders to seed companies and from farmers to policy makers.

Table 3.1 Resistance genes identified in various wheat varieties/lines using molecular markers and/or gene postulation techniques

Study Reference	Lines tested	Lines	Method used	Yr genes tested	Yr genes reported
Tabassum et al., 2010	Commercial varieties	60	Molecular markers	*Yr5, Yr8, Yr9, Yr15* and *Yr18, Yr10*	*Yr9* (17% lines) and *Yr18* (12%)
Qamar et al., 2014	Commercial varieties	52	Molecular markers	*Yr18* and *Sr2*	*Yr18* (25%) and *Sr2* (69%)
Bahri et al., 2011	Commercial varieties	40	Gene postulation	*Yr2, Yr6, Yr7, Yr9, Yr27, YrSU* and *Yr24*	*Yr2* (10%), *Yr6* (10%), *Yr7* (15%), *Yr9* (12%), *Yr27* (22%), *YrSU* (5%) and *Yr24* (2%)
Khan and Ali, 2020	Commercial varieties	35	Molecular markers	*Yr9, Yr17, Yr18, Yr29*, and *Lr46*	*Yr17* (92%), *Lr46* (53%), *Yr18* (14%), *Yr29* (11%), *Yr9* (6%)
Iqbal et al., 2020	Advanced lines	29	Molecular markers	*Yr17, Yr29* and *Yr5*	*Yr17* (91%), *Yr29* (87%) and *Yr5* (50%)
Iftikhar and Ali, 2020	Chinese wheat hybrids	108	Molecular markers	*Lr13, Lr19, Lr26, Lr28, Lr32, Yr5, Yr9, Yr17, Yr18* and *Yr29*	*Lr13* (78%), *Lr19* (71%), *Lr28* (63%), *Lr32* (74%), *Lr36* (66%), *Yr5* (67%), *Yr9* (61%), *Yr17* (57%), *Yr18* (85%) and *Yr29* (80%)

3.5 Virulence Structure of Pakistani *P. striiformis* Population

A high virulence diversity has been reported across Pakistan, based on various studies (Rizwan et al., 2010, Bahri et al., 2011; Ali et al., 2014a; Ali et al., 2017a). The virulence diversity across the country based on the previous studies on virulence phenotyping of *P. striiformis* populations with a total of 223 isolates (Table 3.2). Although the sample size was variable across the studies, it still enabled us to compare the virulence diversity overtime and across provinces.

Analyses of the virulence data revealed substantial variability in virulence diversity across provinces (Fig. 3.3) and over years (Fig. 3.4). The virulence frequencies differed in each province. Virulence Vr1, Vr5, Vr6, Vr7, Vr8, Vr9, Vr17, Vr24, Vr27 and Vrsp were present in AJK, GB, KP and Punjab. Virulence Vr3, Vr4, Vr15, Vr25 and Vr26 were present only in KP. VirulenVr2 was present in GB and KP. Virulence Vr10 and Vr32 was absent in all provinces, while some other virulences were absent in some provinces, e.g., Vr3 and Vr4 in AJK, Gligit Baltistan and Punjab (Fig. 3.3). Interestingly, Vr7 was fixed (100% frequency) in AJK, Gilgit Baltistan and Punjab, also present in high frequency in KP as well, while Vr9 was also in a very high frequency in all provinces. The resistance gene *Yr15* corresponding virulence Vr15 was found only in KP in a very low frequency.

Table 3.2 Details of isolates from various studies used to analyse the diversity in virulence structure across Pakistan

Previous study	Number of isolates virulence phenotyped				
	2004	2006	2010	2014	Total
Rizwan et al., 2010	39	—	—	—	39
Bahri et al., 2011	—	51	—	—	51
Ali et al., 2014a	—	—	127	—	127
Ali et al., 2017a	—	—	—	6	6
Total	39	51	127	6	223

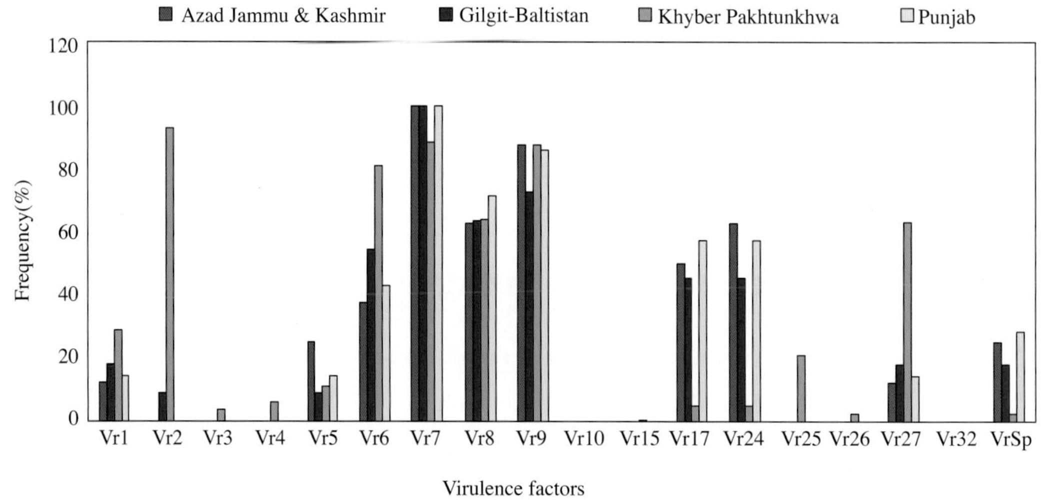

Fig. 3.3 Frequency of different virulence factors across different provinces of Pakistan and based on the previously published studies (Rizwan et al., 2010; Bahri et al., 2011; Ali et al., 2014; Ali et al., 2017)

When considered over years, the virulence frequencies varied among the studied years. Vr1 was present in 2004, 2006, 2010 and 2014. Vr2 was absent in 2004, while it was fixed (with 100% frequency) during 2006, 2010 and 2014. Vr3 was present in 2006 and 2014, while absent during 2004 and 2010. Similarly, the virulence frequency of Vr4, Vr5, Vr15, Vr17, Vr25, Vr26, Vr27 and VrSP ranged from 1% to 83%, while frequencies for Vr2, Vr3, Vr6, Vr7 and Vr9 was fixed in all pathotypes in various years. Virulence frequency of Vr15 was 1% and virulence was observed only in 2010. Virulence to Vr10 and Vr32 was absent in all studied years (Fig. 3.4).

Apart from direct pathotyping studies to track variability in pathogen virulence diversity, the use of trap nurseries based on field exposure of differential lines has been utilized in Pakistan. It provides an indirect idea about the presence of various virulences (Bux et al., 2012). However, the results need to be dealt with more care as the presence or absence of infection under the field condition.

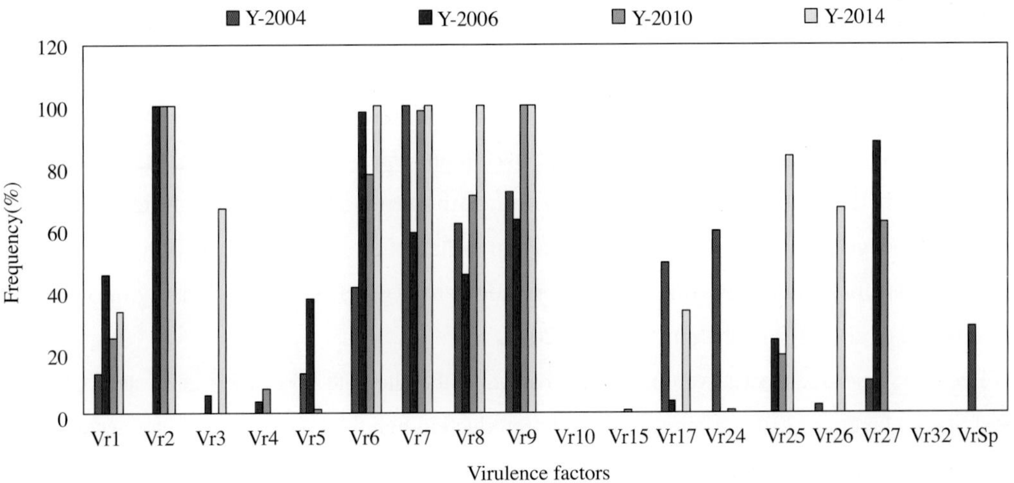

Fig. 3.4 Frequency of different virulence factors across different provinces over years (B) based on the previously published studies (Rizwan et al., 2010; Bahri et al., 2011; Ali et al., 2014c; Ali et al., 2017)

It is important to mention that the virulence diversity is very much high in Pakistan and thus the use of race formula is less useful than providing the exact virulence phenotype (Ali et al., 2014c). Out of the 127 yellow rust pathogen samples phenotyped based on the surveillance effort of 2010, 53 different virulence phenotypes/pathotypes were identified (Ali et al., 2014c), which is much higher than the other worldwide populations (Hovmøller et al., 2016; Ali et al., 2017). These 53 pathotypes included very simple pathotypes with few virulences as well as complex pathotypes with several tested virulences. This could be expected for a population maintained through sexual reproduction, as indicated for the Himalayan region (Ali et al., 2014a; Ali et al., 2010). This information must be considered while deploying resistance genes and devising strategies for disease management.

3.6 Wheat Rust Pathogen Population Structure

Comparative population biology study is important for an efficient disease management strategy at regional level, understanding of accurate population structure of pathogen will enable us to determine how much diversity can be maintained at traits related to pathogenicity, drug resistance or response to abiotic factors. Individuals with such advantageous alleles have an increased potential for invading new host populations or new territories, especially if they have long-distance migration capacity (Brown and Hovmøller, 2002).

The information on pathogen population structure is crucial for genetic improvement of wheat. The biotrophic nature of the pathogen made it difficult to isolate and multiply a large number of samples to do intensive population studies (Bahri et al., 2011). However, the isolation of DNA directly from infected lesion (Ali et al., 2011) and subsequent advances in molecular genotyping (Ali et al., 2017a) and genomic approaches (Bruno-Sancho et al., 2017; Brar et al., 2018) has enabled to make detailed studies on pathogen population structure in different parts of the world, including Pakistan.

Although there is a high diversity in the pathogen population, the selection of few aggressive lineages is also not reported in Pakistan (Khan et al., 2020). This was contrast to other clonal populations, where few lineages are prevalent in a given geographical region (Walter et al., 2016; Ali et al., 2017a; El-Amil et al., 2020).

Recently, *Berberis* species were reported to serve as the alternate host of *Pst* in western China and United States (Jin et al., 2010). Therefore, sexual reproduction may be an important factor accounting for genetic variation in *Pst*, and regions in central Asia may serve as the centers of origin for new races.

Population genetics and genomics tools could also enable to explore the pathogen dispersal and patterns of migrations across Pakistan. Molecular genotyping of samples based on intensive surveillance of grasses, wheat crops, volunteers and *Berberis* spp. would provide a better understanding of the pathogen survival and spread. Such studies should then be extended to neighboring countries to fully explore trans-boundaries migration. This will enable a better disease management strategy not only within Pakistan but at regional level.

3.7 Research Work on *Berberis* and Grasses

The role of sexual reproduction in *P. strifmorms* was suggested based on the sexual reproduction capacity in the Himalayan population, including Pakistan (Ali et al., 2010). The same year in 2010, *Berberis* species were reported to serve as the alternate host of *Pst* (Jin et al., 2010). The pathogen population genetic analyses have also revealed the potential role of recombination in generation of high diversity in Pakistan (Ali et al., 2014b). Additionally, at least 27 species of *Berberis* are present in the Himalayan region of Pakistan and several wheat fields are "protected" by farmers using these Barberry bushes, thus providing a good chance to the pathogen for completion of their sexual cycle (Fig. 3.5). A set of 274 pycnial/aecial-infected *Berberis* leaves and 16 grass samples with uredinial infections were genotyped with EF, ITS region, and β -tubulin primers to assess the presence of various *Puccinia* spp. on these hosts. The samples were collected during the year 2012-2014. Analyses of the data revealed the presence of at least five *Puccinia* spp. viz., *P. brachypodii*, putative *P. coronata-loli* and *P. coronati-agrostis*, *P. striiformis* f. sp. *dactylis* (*P. striiformoides*), and *P. striiformis* on *Berberis* and grasses (Ali et al., 2015). Although further studies are required to identify the exact role of alternate host in generating seasonal variability and disease epidemics onset, there are convincing evidences that the pathogen passes through recombination in the Himalayan region of Pakistan.

Fig. 3.5 *Berberis* plants from the Himalayan region of Pakistan. The close up to the right top shows pycnia on the upper leaf side and aecia on the lower leaf side. The lower three pictures show the use of Barberry as hedges around the wheat fields, providing a good chance to the rust pathogen to complete its sexual cycle.

3.8 Challenges and Opportunities

The higher diversity of pathogen, low variability in crop varieties, utilization of related genetic stock and yellow rust favorable climate during most of the crop seasons remain the major challenges for combating wheat yellow rust disease in Pakistan. This is further complicated by the limited control over seed distribution to avoid large scale spread of the seed of susceptible varieties.

Although the pathogen population is highly diverse, exploitation of new genetic resources could enable to keep the disease below a level to cause economic losses. Future work should be done in collaboration with neighboring countries to exploit new genetic resources and devise a more favorable resistance gene deployment strategy. It will also enable to track the pathogen population at regional level, as suggested earlier (Hovmøller et al., 2016; Ali et al., 2017a), while reducing the impact of invasive strains.

Finally, the replacement of old varieties with the improved resistant varieties should be encouraged on regular basis, instead of few varieties which outperformed in first few years and then continued over decades, despite their susceptibility. This is, however, only possible through the involvement of all stakeholders from breeders to seed companies and from farmers to policy makers.

3.9 Conclusion

Wheat rusts are the major biotic stress for wheat production in Pakistan, with yellow rust being the most important in the northern half of the country. The wheat yellow rust pathogen population is highly diverse in the country with the presence of diverse virulence profiles with virulence to resistance genes which have rarely been exploited. The narrow genetic stock utilized by most of the breeders further complicate the situation. Future work should be done to diversify the wheat genetic stock, while continuing tracking the pathogen population to devise a better disease management strategy.

Acknowledgement

The authors contributing to the chapter were supported by the funds received from the U.S. Department of Agriculture, Agricultural Research Service, under agreement No. 58-0206-0-171 F (Wheat Productivity Enhancement Program, WPEP), ERC funded H2020 Project (GA: 773311 Rust Watch: A European Early-warning System for Wheat Rust Disease) and Start-up Research Grant, Higher Education Commission, Pakistan.

References

Ali S, Gautier A, Leconte M, et al., 2011. A Rapid Genotyping Method for an Obligate Fungal Pathogen, *Puccinia striiformis* f. sp. *tritici*, Based on DNA Extraction From Infected Leaf and Multiplex PCR Genotyping. *BMC Research Notes*, 4: 240.

Ali S, Gladieux P , Rahman H , et al., 2014c. A High Virulence and Pathotypes Diversity of *Puccinia striiformis* f. sp. *tritici* at its Centre of Diversity, the Himalayan Region of Pakistan. *Eur. J. Plant Pathol.*, 140(2): 275-290.

Ali S, Gladieux P, Leconte M, et al., 2014a. Origin, Migration Routes and Worldwide Population Genetic Structure of the Wheat Yellow Rust Pathogen *Puccinia striiformis* f. sp. *tritici*. *PLoS Pathog*, 10(1): e1003903.

Ali S, Gladieux P, Rahman H, et al., 2014b. Inferring the Contribution of Sexual Reproduction, Migration and

Off-season Survival to the Temporal Maintenance of Microbial Populations: a Case Study on the Wheat Fungal Pathogen *Puccinia striiformis* f. sp. *tritici*. *Mol. Ecol,* 23(3): 603-617.

Ali S, Hodson D, 2017b. Wheat Rust Surveillance; Field Disease Scoring and Sample Collection for Phenotyping and Molecular Genotyping. In: Methods in Molecular Biology (ed. Periyannan S).Clifton: Humana Press.

Ali S, Hovmoller MS, Swati ZA, et al., 2015. Diversity of *Puccinia* spp. on Wheat, Grasses and *Berberis* spp. in the Himalayan Center of Diversity of *Puccinia striiformis* f. sp. *tritici*. *BGRI Workshop* 2015.

Ali S, Rodriguez-Algaba J, Thach T, et al., 2017a. Yellow Rust Epidemics Worldwide Were Caused by Pathogen Races From Divergent Genetic Lineages. *Front. Plant Sci*, 8:1058.

Ali S, Shah SJA, Khalil IH, et al., 2009. Partial Resistance to Yellow Rust in Introduced Winter Wheat Germplasm at the North of Pakistan. *Aust. J. Crop Sci,* 3(1): 37-43.

Amil E, Ali S, Bahri B, et al., 2020. Pathotype Diversification of the Invasive PstS2 Clonal Lineage of *Puccinia striiformis* f. sp. *tritici* Causing Yellow rust on Durum and Bread Wheat in Lebanon and Syria. *Plant Pathol,* 69: 618-630.

Arif M, Khan MA, Akbar H, et al., 2006. Prospects of Wheat as a Dual-purpose Crop and its Impact on Weeds. *Pak. J. Weed Sci. Res*., 12:13-18.

Bahri B, Shah SJA, Hussain S, et al., 2011. Genetic Diversity of the Wheat Yellow Rust Population in Pakistan and its Relationship with Host Resistance. *Plant Pathology*, 60(4): 649-660.

Beddow JM, Pardey PG, Chai Y, et al., 2015. Research Investment Implications of Shifts in the Global Geography of Wheat Yellow Rust. *Nature Plants*, 1(10): 1-5.

Brar GS, Ali S, Qutob D, et al., 2018. Genome Re-sequencing and Simple Sequence Repeat Markers Reveal the Existence of Divergent Lineages in the Canadian *Puccinia striiformis* f. sp. *Tritici* Population with Extensive DNA Methylation. *Environ. Microbiol*., 20:1498-1515.

Brown JK, , Hovmøller MS, 2002. Aerial Dispersal of Pathogens on the Global and Continental Scales and its Impact on Plant Disease. *Science*, 297(5581): 537-541.

Bueno-Sancho V, Persoons A, Hubbard A, et al., 2018. Pathogenomic Analysis of Wheat Yellow Rust Lineages Detects Seasonal Variation and Host Specificity. *Genome Biology and Evolution*, 9: 3282-3296.

Bux H, Ashraf M, Chen X, et al., 2011. Effective Genes for Resistance to Yellow Rust and Virulence of *Puccinia striiformis* f. sp. *tritici* in Pakistan. *Afr. J. Biotechnol,* 10(28): 5489-5495.

Bux H, Ashraf M, Husain A, et al., 2012. Characterization of Wheat Germplasm for Yellow Rust (*Puccini striiformis* f. sp. *tritici*) Resistance. *Aust. J. Crop Sci,* 6 (1):116-120.

Chen W, Wellings C, Chen X, et al., 2014. Wheat Yellow (yellow) Rust Caused by *Puccinia striiformis* f. sp. *tritici*. *Mol. Plant Pathol,* 15(5): 433-446.

de Vallavieille-Pope C, Ali S , Leconte M, et al., 2012. Virulence Dynamics and Regional Structuring of *Puccinia striiformis* f. sp. *tritici* in France Between 1984 and 2009. *Plant Dis*, 96: 131-140.

Dean R, Van-Kan JA, Pretorius ZA, et al., 2012. The Top 10 Fungal Pathogens in Molecular Plant Pathology.

Mol. Plant Pathol, 13: 414-430.

Hovmøller MS, Walter S, Bayles R, et al., 2016. Replacement of the European Wheat Yellow Rust Population by New Races from the Centre of Diversity in the Near-Himalayan Region. *Plant Pathol*, 65:402-411.

Hovmøller MS, Walter S, Justesen AF., 2010. Escalating Threat of Wheat Rusts. *Science*, 329:369.

Iqbal A, Khan MR, Ismail M, et al., 2020. Molecular and Field-based Characterization of Yellow Rust Resistance in Exotic Wheat Germplasm. *Pak. J. Agric. Sci*, 57:1457-1467.

Jin Y, Szabo LJ, , Carson M., 2010. Century-old mystery of *Puccinia striiformis* life History Solved with the Identification of *Berberis* as an Alternate Host. *Phytopathology*, 100(5): 432-435.

Khan MR, Imtiaz M, Ahmad S, et al., 2019. Northern Himalayan Region of Pakistan with Cold and Wet Climate Favours a High Prevalence of Wheat Powdery Mildew. *Sarhad J. Agric,* 35:187-193.

Khan MR, Imtiaz M, Ahmed B, et al., 2020. Diversity in *P. striiformis* Populations Causing the 2013 Yellow Rust Epidemics on Major Wheat Cultivars of Pakistan. *Mycologia*, 15: 871-879.

Khan MR, Rahman ZU, Nazir SN, et al., 2019 Genetic Divergence and Diversity in Himalayan *Puccinia striiformis* Populations from Bhutan, Nepal and Pakistan. *Phytopathology*, 109:1793-1800.

Kisana SN, Mujahid YM, Mustafa ZS, 2003. Wheat Production and Productivity 2002-2003. A Technical Report to Apprise the Issues and Future Strategies. Published by Coordinated Wheat, Barley and Triticale Program, National Agricultural Research Centre, Pakistan Agricultural Research Council, Islamabad.

Line RF, 2002. Yellow Rust of Wheat and Barley in North America: a Retrospective Historical Review. *Annu. Rev. Phytopathol*, 40(1): 75 -118.

Rizwan S, Iftikhar A, Kazi AM, et al., 2010. Virulence Variation of *Puccinia striiformis* Westend. f. sp. *tritici* in Pakistan. *Arch. Phytopathol. Pflanzenschutz*, 43(9): 875-882.

Roelfs AP, Bushnell WR, 1985. The Cereal Rusts (Vol. 2, p. 606). Orlando, FL: Academic Press.

Saari EE, Hashmi NI, Kisana NS., 1995. Wheat and Pakistan, an update (*Yr* 95 doc.).

Sinha P, Chen X, 2021. Potential Infection Risks of the Wheat Yellow Rust and Stem Rust Pathogens on Barberry in Asia and Southeastern Europe. *Plants* (Basel), 10(5):957. DOI: 10.3390/plants10050957.

Sobia T, Muhammad A, Chen X, 2010. Evaluation of Pakistan Wheat Germplasm for Yellow Rust Resistance Using Molecular Markers. *Sci. China Life Sci*, 53(9): 1123-1134.

Walter S, Ali S, Kemen E, et al., 2016. Molecular Markers for Tracking the Origin and Worldwide Distribution of Invasive Strains of *Puccinia striiformis*. *Ecol. Evol*, 6: 2790-2804.

Chapter 4

Wheat Yellow Rust in Iran – Status, Challenges and Perspectives

Ahamd Abbasi Moghaddam

Seed and Plant Improvement Institute, Agricultural Research, Education and Extension Organization (AREEO), Karaj, Iran

Email: abasimoghadam@gmail.com

Abstract: Wheat yellow (stripe) rust is the major disease of wheat in most part of Iran, during winter or early spring, especially at higher elevations. The causal pathogen on *Triticeae* is *Puccinia striiformis as* revealed based on the molecular and morphological study of specimens collected from different hosts in Iran. This chapter describes the main agro-ecological traits of Iran, the economic importance of wheat yellow rust and historical epidemics. An effort is made to summarize the results of the work done on wheat yellow rust epidemiology, resistance gene in host and its mechanism, pathogen population biology, population genetics and physiological races identification. It concludes on the integration of this information for the management of wheat yellow rust resistance and future avenues to work in Iran.

Keywords: Wheat Yellow Rust; Resistance; Pathogenicity

4.1 Main Agro-ecological Traits of Iran

The Islamic Republic of Iran covers 164.8 million hectares between latitudes 25° and 40° N in the Middle East. Iran has a 432 km border with Azerbaijan and 35 km with Armenia to the north-west; 992 km border with Turkmenistan to the north-east; 909 km border with Pakistan and 936 km border with Afghanistan to the east; 499 km border with Turkey and 1,458 km border with Iraq to the west; 1,000 km border at the Caspian Sea to the north; and 3,200 km border along the waters of the Persian Gulf and the Sea of Oman to the south. Iran is the eighteenth largest country in the world and the second largest in the Middle East.

Iran is considered predominantly dry due to the location in the arid and semi-arid region of the world and its geographic features. Thus, the country receives an average annual rainfall of 240 mm, less than a third of world average precipitation. However, annual precipitation in the inland dry deserts of the country can be as low as only 10 mm. As a result, most rivers are seasonal and their flows depend heavily upon the amount of rainfall. The altitude range varies from −26 m below the sea level to 5,770 m above the sea level. However, the main part of the country is located on highlands with more than 1,200 m above the sea level. The temperature range changes from the minimum of −35 ℃ in the North West of the country to maximum of 50℃ in Persian Gulf coasts in the South.

Out of 164.8 million hectares (ha), land of Iran, 86 million ha (52.4%) are rangelands, 14.2 million ha (8.6%) are forests and 32 million ha (19.5%) are deserts including bare salty lands. Approximately only 18.5 million ha (11%) are under cultivation, of which 8.5 million ha are irrigated and 10 million ha are rainfed.

Existence of various cultures and ethnicities, varied climatic conditions and long history of agriculture has contributed to the richness and uniqueness of biodiversity in the country. Because of its large size and varied ecosystems, Iran is one of the most important countrics for conservation of biological diversity in the Middle East and West Asia.

Iran's plateau with a vast desert located in the central areas, and two mountain ranges, Zagros in the west and Alborz in the north, comprise a significant portion of its territory. Iran's topography has given rise to four main ecological (floristic) zones or biomes, namely: Irano-Touranian arid and semi-arid zone with two plain and mountain sub-regions, Zagrosian, Hyrcanian, and Khalij-o-Omanian zones. Most of Iran's forest covers are located in Hyrcanian, Arasbaranian and Zagrosian zones. The land areas of Iran's biomes in the country are shown in Table 4.1.

Table 4.1 The land area coverage for prevalent biomes of Iran

Biome	Area (ha)
Irano-Touranian plain and mountain Area	3,452,775
Zagrosian	4,749,000
Hyrcanian	2,800,000
Khalijo-Ommanian	2,130,000

Iranian agriculture is thousands of years old and this reflects the length of time during which soil and water resources of the country have been utilized for various crop production systems. Both irrigated and rainfed farming (dry farming) are practiced in different parts of the country while the area devoted to each system varies considerably depending on the annual precipitation and agro-climatically conditions. Rainfed agriculture and dry farming are most successful in the north, west and northwestern of Iran. In other parts of the country, dry farming is also practiced in hilly areas, but the yields are very low. In the central plateau, as well as the southern plains and the southern coastal areas of Iran, crop production is mostly possible only under irrigation. This is because of low rainfall and high evaporation rates. All agricultural lands are owned by the people, except some portions of which are used by the government sector for particular (research, development projects, etc.) purposes.

About 91% of the irrigated land is under annual crops (The remaining 9% is used for production of perennial crops, mostly fruit trees). In rainfed areas, annual crops constitute about 98% of the total production. During dry years, about 8% to 12% of the total production comes from dry land areas. However, in wet years this figure can rise to 35%.

Traditional small-scale farming was the main structure of farming communities for centuries and this has caused a great diversity in crop production and tremendous accumulation of indigenous knowledge in farming practices and food production. Therefore, natural farming used to be a widespread practice not more than half a century ago and is still appreciated to some extent by the farmers producing under marginal or unfavorable conditions or by the consumers looking for tasty, safe and high-quality products. Therefore, the traditional systems of food production are almost organic both in technical and social terms, but they are in operation in small scale farming or in remote areas. Thus, there is considerable land area under organic management, but only a small number of these farms are registered as organic farms.

Most of Iran is located in the Palaearctic realm and is considered the centre of origin of many genetic resources of the world, including many of the land races and wild relatives of commercially valuable crop plants such as wheat, food and forage legumes, fruit trees, or medicinal and aromatic species. Scientific studies denoted that 8,000 to 10,000 years ago, probably in the Fertile Crescent, in a region that nowadays comprises Northern Iran, the hybridization between *Triticum turgidum* and *Aegilops tauschii* gave rise to the hexaploid *T. aestivum*. That after domestication and centuries of cultivation and selection, resulted in the bread wheat that is cultivated today (Dubcovsky and Dvorak, 2007; El Baidouri et al., 2017; Venske et al., 2019). According to geographical situation of Iran, wheat became the main food plant and staple crop for food security of Iran. Around 6 million hectares of Iran agricultural field occupied by wheat cultivation during recent years. Wheat was cultivated around 5.8 million hectares during 2020 in Iran from which 1.85 million hectares was irrigated and 3.95 million hectares was rainfed (https://www.irna.ir/news/83613412).

In this chapter the results of research on the economic importance of wheat yellow rust disease that was a historical constrains in wheat production, epidemiology of wheat yellow rust, resistance, resistance gene of wheat cultivars, and resistance mechanism, pathogen population biology, population genetics, and pathogenicity mechanism, and integrated pest management (IPM) in Iran has been discussed.

4.2 The Economic Importance of Wheat Yellow Rust and Historical Epidemics

Wheat yellow (yellow) rust is the major disease of wheat in most part of Iran, especially at higher elevations during winter or early spring. The causal pathogen on *Triticeae* is *Puccinia striiformis* s. str., species based on the molecular and morphological data study of specimens collected from different hosts in Iran (Abbasi et al., 2005). The *Puccinia striiformis* Westend. f. sp. *tritici* Erikss. (*Pst*) is accepted as the wheat yellow rust pathogen. This fungal pathogen belongs to *Basidiomycota* division, *Pucciniomycetes* class, *Pucciniales* order, *Pucciniaceae* family and *Puccinia* genus. It frequently causes epidemics resulting in a significant wheat yield losses and a reduction in the quality of the grain and forage. Cultivation of susceptible wheat cultivars under favorable climate condition for *Pst* expansion and rapid development of new races of *Pst*, had caused a lot of food scares during history.

Yellow rust damaged wheat, producing low vigor seeds which if germinate produced poor emergence that impact on next growing season production. During centuries of yellow rust, wheat interaction, the boom and bust cycle of disease and host was occurred. The selection of healthy resistant plants and exchange of healthy seeds was practiced by farmers to ensure yield under stress condition. This resulted in maintaining diversity for the resistance in the wheat local landraces. The ability of yield production under stress is so important that many farmers on west of Iran used to grow the Sardary local landrace under rain-fed condition. It is interesting to know that the Sardary local landrace was a mixture of a lot of wheat morphotype with different construction of slow rusting resistance genes that despite the epidemic of a *Pst* could produce some yield.

Effect of different parameters on wheat with yellow rust interaction like susceptibility of the cultivars, virulence of *Pst* races, earliness of the initial infection, rate of disease development, and duration of disease incidence on the amount of yield losses has been reported (Afzal et al., 2007; Ahmad et al., 2010; Chen, 2005). Study on the effect of yellow rust among different type of wheat cultivars revealed 4%~66% loss in Iran. The mean thousand kernels weight (TKW) losses of 41% were revealed on susceptible cultivars, 7.6% on cultivars with slow rusting gene(s) and 4.4% on cultivars with effective race-specific gene(s). The yield losses were 65.6% on susceptible cultivars, 15.9% on cultivars with slow rusting gene(s) and 7.3% on cultivars with effective race-specific gene(s) (Safavi, 2015).

Historically several epidemics of yellow rust disease had been occurred and the boom and bust cycles of wheat with yellow rust lead to adoption of diverse local landraces which were mixture of several wheat genotypes with different resistant genes over different location in Iran. Starting scientific plant breeding on cereals after 1930 with selection on local landraces attract the attention of scientists for the genetic improvement against pest and disease. Cereal rusts including the wheat yellow rust was first reported from Iran in 1946 (Esfandiari, 1947). It was also the first scientific report of 4.5% disease loss on wheat yield of Iran by yellow rust. Another epidemic of *Pst* was reported on wheat cultivation at the north of Iran during 1967 with losses of 5.4% (Niemann et al., 1968). Later several yellow rust epidemics occurrence was reported every three or four years in Iran. Scientists estimated that overall losses in such years might be as high as 30%~40% (Khazra and Bamdadian, 1974; Bamdadian and Torabi, 1978; Torabi, 1980). Due to population, growth and need for high yield wheat, cultivars such as Azadi, Quds and Falat, released in Iran and recommended for cultivation. At their release, these were resistant but became susceptible to yellow rust after a few years. Cultivation of Falat cultivar which had *Yr9* and the emergence of new pathotypes of yellow rust with virulence against *Yr9* gene were the most important factors for epidemics of yellow rust during 1991-1995 in many parts of the wheat growing areas of Iran (Bamdadian et al., 1991; Torabi et al., 1995; Bimb and Johnson, 1997). The yield losses caused by the disease were estimated to have been about 1.5 million tons in 1993 and 1 million tons in 1995 (Torabi et al., 1995). This epidemic triggered the establishment of cereal pathology unit of cereals research department in the Seed and Plant Improvement Institute under Agricultural Research Education and Extension Organization, the deputy of Iran Ministry of agriculture Jahad. The mission of cereal pathology unit demarcated for regular monitoring of cereal disease prevalence, study on their pathogen population diversity and evaluation of commercial cultivars reaction to prevalent disease pathotypes especially yellow rust. Since then, detail of monitoring of wheat cultivation, disease incidence and evaluation of new commercial cultivars reports are available and every outbreak has been reported. Then the epidemic of new *Pst* race virulent on Chamran cultivar that caused effectiveness loss of resistant gene *Yr27* from Kermanshah and Fars provinces reported in 2003 (Afshari et al., 2004). Boyerahmad and Dena are two cold areas of Kohghilooye and Boyerahmad province where bread wheat is cultivated as main crop. Severe epidemic was observed in Dashteroom region of Kohghilooye and Boyerahmad province where 100% of fields were infected in 2005 (Viani, 2009). Global study for yellow rust status during 2000-2009 revealed localized epidemics incidence occurrence in 2 out of 5 years and 1%~5% crop losses in some seasons for over 25% of wheat growing area of Iran (Wellings, 2011). Mild winter and adequate rainfall favoring rust development in

2010 reported to be the cause for early yellow rust outbreak. In order to manage the disease and control the loss, over 750,000 ha were sprayed against yellow rust in Iran (Solh et al., 2012). Later there was report of yellow rust incidence in some parts of Iran in 2011, but the crop loss due to this disease was not significant. But it is important to notice that even in the absence of yellow rust epidemic, some crop losses due to the susceptibility of some commercial cultivars to limited infection has been reported (Ghaffari and JalalKamali, 2013).

According to recent scientific finding, the lack of predictability of invasive races in terms of their origin and adaptability found to be the result of the emergence and prevalence of races of few divergent lineages from diverse populations (Ali et al., 2014). Iran is also close to the pathogen centre of diversity and potential center of origin, i.e., the Himalayan region (Ali et al., 2014; Ali et al., 2018) and thus regular invasions could enter Iran from these areas. In order to prevent the future epidemic of yellow rust pathogen scientists placed emphasis on the need for the efficient sharing of knowledge, germplasm, rust diagnostic facilities and information at national, regional and continental scales (Ali et al., 2017). The outcome of climate change that presents with warmer winters could be encourage earlier yellow rust infection and spread in the cold regions of Iran. In the warmer regions the emergence of more aggressive races of yellow rust better adapted to warmer temperatures base on the reports of such incidence in the last two decades (Chen, 2005; Khanfri et al., 2018), could also lead to massive epidemics.

4.3 Wheat Yellow Rust Epidemiology

It is necessary to know the climatic condition of Iran as a prerequisite to understanding the yellow rust epidemiology, as the disease is present in diverse climatic zones, including the areas with the presence of alternate host (Fig. 4.1). Three main climatic zones are Arid and semi-arid in the interior and far south, Mediterranean climate mainly in the western Zagros Mountains, the high plateau of Azerbaijan, and the Alborz mountains and Temperate humid and semi-humid mainly in the Caspian Sea coasts, but also in West Azerbaijan and the South Western Zagros.

Winter, facultative and spring wheat types, are grown in different agro-climatic zones under irrigated and rainfed conditions. The temperate zone considered promising area for wheat production and high grain yields have been recorded at Kangavar in Kermanshah province (14 t/ha) and in Daryoun in Fars Province (12 t/ha) which are part of this agro-climatic zone (Ghaffari and JalalKamali, 2013). Initial reports of yellow rust epidemiological study in Iran brought attention to summer conditions of many wheat growing areas of Iran, that are typically dry and hot. Researcher tried to detect the mean of yellow rust survival during summer time. It has been reported that in regions with moderate and cold climate, temperature is mild during

Fig. 4.1 Wheat and barberry in the field and infected leaves in Iran
a: Wheat yellow rust resistance evaluation in a hot spot, b: Yellow rust pustule on wheat susceptible cultivars, c: Barberry as alternate host, d & e: Infected leaves of barberry

summer and yellow rust can survive on volunteer wheat and wild grasses in the absence of a wheat crop in fields (Bamdadian and Torabi, 1978; Torabi, 1980, 2004). Wheat yellow rust spread in the field highly correlated to the amount and duration of rainfall, temperature, and relative humidity during the growing season. High correlation of primary infections observation on location understudy with the mean daily temperature in March and April has been reported according to four years studies on the provinces of Ardebil, Moghan, Fars, Golestan, Hamedan, Khuzestan, West Azarbaijan, Mazandaran, Kermanshah, Sistan and Baluchistan of Iran (Torabi, 2004). Torabi believed that primary infection and disease development during March to June in Iran were strongly correlated with some factors. First were weather parameters, the second was the amount of viable inoculum on volunteer wheat and grass hosts, and the third were latent infections of seedlings in the new crop during the autumn and winter months. Another factor was over-wintering latent infections of yellow rust from October to January on the first and second leaves of seedlings of Falat in Mazandaran and west Azarbaijan provinces of Iran (Torabi, 2004).

Urediniospore germination of different *Pst* isolates at various temperatures on agar and wheat detached leaves assessed in Iran and their performance over a range of temperatures found inherently different despite the similarity of optimum temperature for *Pst* urediniospores germination (Elahinia, 2000).

Both additive and dominance gene action was found to be involved in the genetic control of the latent period of the *Pst* on wheat in Iran. Mostly high narrow-sense heritability of latent period stated as importance of additive effects of genes in governing this character and the possibility of selection for a longer latent period in breeding programs was encouraged (Hamid Dehghani and Moghaddam, 2004).

Yellow rust pathogen complete life cycle on barberry plants by basidiospores of the *Pst* and production of aeciospores under controlled conditions (Jin et al., 2010). So far natural infection of *Pst* has been detected in China. Sinha and Chen reported a high probability of *P. striiformis* infection on barberry in central and western Asia surrounding the Mediterranean Sea and the Caspian Sea regions, a part of Iran (Zhao et al., 2016; Chen, 2020).

There is a report of observing the pathogenicity of yellow rust isolates collected from different species of *Hordeum spontaneum*, *H. glaucum*, *Aegilops crassa*, *Ae. cylindric*, *Ae. triuncialis*, *Ae tauschii*, *Ae. squarrosa*, *Agropyron repens*, *Bromus tectorum*, *B. scoparius* and *Phalaris minor* on wheat in Iran. The presence of active yellow rust infections on wild grasses and wheat in the highlands in late-harvested fields up to September was described. Transition to the new wheat crop thought to be the cause of primary infections and claimed that infections could remain latent until the establishment of favorable conditions (Torabi, 2004).

Distribution and genetic diversity of aecial infection on barberry bushes and their relation to wheat yellow rust was studied in the west of Iran. Samples of infected leaves to aecial stage on barberry bushes were collected from Borujerd, Dorud, Azna and Aligudarz regions. To identify the rust species, in vivo and in vitro pathogenicity test were conducted by artificial inoculation of wheat seedlings using collected aeciospores from each region. The results showed the production of urediniospores of *Pst* in inoculated wheat seedlings (Mehdinia et al., 2016). Molecular detection and sequencing of the rDNA regions of representative aecial isolates on barberry were done. Genetic diversity of 134 aecial isolates was also studied by amplification of IGS1-rDNA region. The Results also confirmed the pathogenicity test and showed that barberry could be the aecial host of yellow rusts in Lorestan province, west of Iran. The IGS-rDNA amplification of aecial isolates from Azna, Aligudarz, Dorud and Broujerd showed a genetic variation among aecial isolates collected from sampling regions in Lorestan province, Iran. *Berberis vulgaris* reported to be most prevalent over high-land of Dorud and Azna and *Berberis integerrima* over high-land of Aligudarz and Broujerd (Mehdinia et al., 2016).

Scientists argued that in a condition that the alternate host provides the main source of inoculum needs to be eradicated otherwise, eradication might have less or any effect on *Pst* disease management. Therefore, they suggested future surveillance of yellow rust pathogens on both alternate hosts and cereal crops is required to identify new virulence genes and races to acquire a complete understanding of the role of alternate hosts (Zhao et al., 2016).

4.4 Resistance, Resistance Gene of Wheat Cultivars, and Resistance Mechanism

Wheat landraces are valuable sources of resistance to stresses; those developed over long periods and have complex resistance in their geographical regions. Northwestern Iran is located in the Fertile Crescent, presents an important wheat and pathogens co-epicenter. It seems that the two type of race-specific resistance and race-non-specific evolved through a history of boom and bust cycle between pathogen and host there. Identification and utilization of different types of resistance is an important strategy in management of wheat yellow rust and there are many studies on survey of different cultivars trying to identify the resistant gene related to that type of reaction. Although the different type of resistance from different point of study has been mentioned, it seems that the resistance of cultivars to *Pst* could be studied under two categories of the all-stage resistance that is similar to race-specific resistance and the adult plant resistance that is similar to nonrace-specific resistance.

The ability of some wheat cultivars for resistance to yellow rust infection has been known for many years, as farmers prefer the landraces that less infected to yellow rust. During the recent years, result of first investigation on the inheritance of resistance to *P. striiformis* that was made by Biffen during 1905 showed that the resistance of Rivet wheat (*Triticum turgidum*) was controlled by a single recessive gene. That established the new era of scientific investigation on the resistant gene discovery among land races, commercial cultivars and their wild relatives (Biffen, 1905).

Later scientists found that some resistant cultivars are deemed to be non-specific, but changes in the pathogen diversity have caused the failure of many resistant cultivars to yellow rust, therefor they suggesting race specificity (Roelfs et al., 1992). Change in *Pst* races ratio during the interaction with commercial resistant cultivars brought non-durability of resistance in cultivars. That promote breeders to look for other types like slow-rusting and non-race-specific resistance genes. It has been mentioned that non-race-specific resistance employs two genetic arrangements. One is on the base of gene-for-gene that triggered when the host and the pathogen genes have some interaction, the others chemical or structural features that affect the rate of epidemic development before or after *Pst* penetration to resistant cultivars. Slow-rusting wheat cultivars infected with *Pst* exhibit, longer latent period, smaller and fewer uredinia, and less spore production than susceptible cultivars (Parlevliet, 1985; Roelfs et al., 1992). Latest report for yellow rust resistant genes indicated so far; 83 officially named genes, 67 temporarily designated resistance genes, and over 300 quantitative trait loci (QTL) have been described in the world (Chen, 2020).

Start of modern plant breeding after 1930 in Iran and introduction of commercial cultivars by selection on the local landrace and extensive sowing of cultivars like Shahpasand bring attention to yellow rust incidence and later yield loss on wheat caused by yellow rust (Niemann et al., 1968). Then change of the cultivar's reaction to susceptible with change of the *Pst* physiologic race frequency in Iran reported. Later the need for high yield wheat cultivar due to population growth increased and cultivars such as Azadi, Quds and Falat became popular. Extensive cultivation of Falat cultivar which had *Yr9* on large area prepared selection pressure on the *Pst* pathogen and new virulent physiologic race emerged (Bamdadian et al., 1991). After that incidence of yellow rust epidemic advanced the evaluation of breeding material for resistance to yellow rust races in different agroclimatic region of the Iran and scientific reports of cultivar evaluation for resistance started (Torabi et al., 1995). Johnson reported a severe epidemic on cultivar Seri82 (given the name Falat, in Iran) during his visit in 1995 and cooperate with Bimb to evaluate 280 wheat lines in which 128 were derived from the CIMMYT 25th International Bread Wheat Screening Nursery and reported the present of substantial number of susceptibilities to yellow rust (Bimb and Johnson, 1997). Evaluation of the 23 advanced, promising and commercial bread wheat cultivars resistance present different reactions but two cultivars, M-70-4 and MV-17 reported resistant to five tested pathotypes at seedling and adult plant at four locations (Torabi and Nazari, 1997).

During field evaluation of breeding material, all Iranian old commercial wheat varieties, which had been released before 1992, reported to become highly susceptible and virulence on many newly released varieties such as Tajan, Mahdavi, Alamout, Zarrin, Alvand, Chamran and wheat advanced lines C-73-5, 8-75-20, M-75-10, M-75-19 and N-75-5 has been described (Nazari et al., 2001). Later evaluation of 56 advanced wheat lines/varieties from CIMMYT for resistance to yellow rust at seedling stage in greenhouse and at adult-plant stage in the field reported by Nazari et al. in 2001. Three pathotypes 6E0A, 70E16A+, and 134E134A+ were used in greenhouse and pathotype 134E134A+ in field. While all genotypes were completely resistant to pathotypes 6E0A and 70E16A+ at seedling, only line CHIL/BUC reported completely resistant and three lines showed intermediate resistance to pathotype I34E134A+, the rest were susceptible. They considered genotypes showing susceptible or intermediate infection types at seedling stage, but less than 30% rAUDPC in the field, as adult-plant resistant (Nazari et al., 2001).

Diallel analysis of infection types on four advanced resistant lines of wheat indicated that line M-78-16 had the greatest general combining ability to decrease infection type and therefore it was recommended for utilization in breeding resistant cultivars to yellow rust (Zahravi et al., 2004). Then result of study in a half-diallel design using six wheat cultivars of; Tiritea (a susceptible check), Tancred, Kotare, Otane, Karamu and Briscard by three yellow rust pathotypes 7EI8A− , 38E0A+ and 134E134A+ brought recommendation for Briscard and Kotare to be used in breeding programs for development of resistant cultivars to yellow rust (Dehghani et al., 2005). Then reaction of wheat resistant and susceptible cultivars to yellow rust races 70E128 and 14E14 studied. Urediniospore germination in all genotypes/isolates/leaf stage combination described as high but as plant aged disease latent period increased that resistant cultivars with longer latent periods reported (Elahinia and Tewari, 2005).

During evaluation of promising wheat genotypes for cold zone of Ardabil province, the reaction of 20 genotypes to yellow rust also described. The lines of C-80-3, C-80-4, C-80-5, C-80-11, C-80-12, C-80-16 and C-80-20 were reported highly resistant (Safavi et al., 2008). Also evaluation of 200 rainfed wheat advanced lines and cultivars for resistance to yellow rust indicated that 39% material were completely resistant, and 7% moderately resistant to race 134E134A+ but 30% were completely resistant, and 9% moderately resistant to race 134E142A+ (Malihipour and Torabi, 2008).

Components of resistance to yellow rust in 47 wheat genotypes of national plant gene bank of Iran assessed by four pathotypes of 6E6A+, 134E134A+, 134E142A+ and 6E4A+ at seedling stage. Correlation Analysis among components of resistance showed that association between latent period and pustule density was not influenced by the pathotype but relationship between latent period and pustule size was affected by the pathotypes of the pathogen. Latent period described with the high effect on the variation of related principal components. The results of this investigation demonstrated that latent period could be a suitable replacement for other components of resistance to yellow rust in order to select superior genotypes (Zahravi et al., 2009). Combining the ability of latent period of yellow rust in advanced lines of wheat practiced by estimation gene effects and the importance of both additive and non-additive gene effects on them (Zahravi and Bihamta, 2010). The resistance of 4096 bread wheat accessions to yellow rust from national plant gene-bank of Iran at the field of Gharakhil, Sari monitored and reaction of 1226 accessions reported with different level of resistance to local races (Arshad et al., 2012).

The resistance to yellow rust in 72 wheat genotypes from collection of National Plant Gene Bank of Iran was evaluated at two hot spots of Sari and Karaj, Iran. Each component of partial resistance to yellow rust reported important. The genotypes Cluster analysis placed accessions in two groups of susceptible and partially resistant (Zahravi et al., 2012). Then 45 advanced lines of wheat along yellow rust differential set and near isogenic lines evaluated for resistance by races collected from Ardebil, Zarghan and Sari. Yellow rust races 6E6A+, 6E150A+ and 198 E150A+ determined and lines No. 16, 18, 20, 22, 29, 30, 31, 40, 41, 42, 43 and 44 exhibited complete resistance to all races. The presence of all stage resistant genes like *Yr1, Yr3, Yr4, Yr5, Yr10, Yr15, Yr24, Yr26, YrSP, YrND, YrSD, YrCV* or other unknown gene/s in these lines assumed (Omrani et al. 2013).

Bread wheat cultivars and advance lines of Pishtaz, MV17, Moghan 3, SERI/KAUZ, KAUZ//KAUZ/PVZ and Bolani were crossed using half diallel to determine the heritability, gene action and estimate of genetics parameters for yellow rust resistance component. Results indicated that latent period, infection type, pustule size, and number of pustules were recessive in some genotypes but dominant in others. Broad and narrow sense heritability for all four traits were relatively high, therefore, selection based on these traits for enhancement of resistance reported useful (Khodarahmi et al., 2014). The effectiveness of different types of resistance was compared in field plots at Ardabil Agricultural Research Station (Iran) during 2011-2013. Characters of yield and yield components along with slow rusting parameters including final rust severity (FRS), apparent infection rate (r), relative area under disease progress curve (rAUDPC) and coefficient of infection (CI) for 16 wheat cultivars reported. The cultivars Zareh, Bezostaya, Morvarid, Sisons and Gonbad, with genotypes Chamran and Rasad reported to have genes for varying degrees of slow rusting (Safavi, 2015).

Then evaluation of cultivars characters by molecular markers as rapid with high precision method started. The 124 Iranian wheat genotypes were characterized for the linkage groups *Lr19/Sr25, Lr35/Sr39, Lr24/Sr24, Lr20/Sr15, Sr31/Lr26/Yr9, Lr37/Sr38/Yr17* and the widely used gene combination of *Lr34/Yr18*. The most frequent detected genes were *Lr26/Sr31/Yr9* and *Lr34/Yr18* in 40 and 39 genotypes, respectively (Dadrezaei and Nazari, 2015).

To determine partial resistance to yellow rust, 21 bread wheat lines of cold climate zone of Iran were studied in the field condition of Ardebi. Lines C-91-1, C-91-7, C-91-8, C-91-15 and C-91-18 described resistant or had low infection at both seedling and adult plant stages. Lines C-91-12 and C-91-21 identified susceptible at seedling stage, but resistance at adult plant stage. These lines reported to have slow rusting (partial) resistance. Lines C-91-13 and C-91-14 recognized moderately resistant at seedling and resistant at adult plant stage, while lines C-91-3, C-91-4, C-91-16 and C-91-17 reported with moderate level of partial resistance (Ajirloo et al., 2016).

Reaction of 150 doubled haploid lines with parents and check cultivars to yellow rust was reported at seedling and adult plant stages in Karaj. Out of them 28 doubled haploid lines at both stages of seedling and adult plant reported resistant to local race of Karaj. The probability of carrying one or more minor adult plant resistance genes in addition to seedling major genes reported (Bakhtiar et al., 2016).

Selected wheat fields of Fars province: including Zarghan, Marvdashat, Fassa, Mammassani, Darab and Eghlid monitored for reaction to *Pst* during 2003-2013. Some cultivars including Chamran, Shirodi, Shiraz, Darab 2, Falat, Star, kavir, Zarin, Shahriar, Alvand, Alemot and Bahar became susceptible to yellow rust prevalent race, during or before the last decade. Cultivars including Yavarous, Zareh, Mihan, Siravan, Chamran 2, Ophogh, Behrang and Shabrang have conferred moderately to intermediate resistance reaction to the yellow rust disease. The reaction of cultivars Sivand, Parsi, Pishgham, Marvdasht and Aflak observed in range of moderately resistance to moderately susceptible. The response of Pishtaz varied from moderately susceptible to resistance. High level of resistance reported since the time of Nicknejad introduction. Some scientists believed that changes in yellow rust races frequency, changes in the resistance of some of the cultivars can be attributed to weather conditions and environmental factors (Zakeri et al. 2017).

Reaction of 20 wheat cultivars to yellow rust was evaluated at adult plant stages in Nishabour (north east of Iran). Cultivars Tajan, Nishabour, Ofog, Chamran and Shirodi reported more tolerant to *Pst* infections (Dadrezaei et al. 2018). During same year 41 genotypes of bread wheat collection of national plant gene bank of Iran evaluated with four races of yellow rust at seedling stages. The *Pst* races has been collected from Gorgan, Karaj, Mashhad and Sari and identified as 6E6A+, 134E134A+, 134E142A+ and 6E4A+ respectively. To all four races, 29 genotypes reported resistant and there was not any virulence on *Yr1, Yr3, Yr4, Yr5, Yr8, Yr10, YrSP, YrSU, YrSD, YrCV* and the potential for resistance to yellow rust in bread wheat collection of national plant gene bank of Iran been reported (Zahravi and Afshari, 2018).

Reaction of 284 accessions from bread wheat were evaluated at adult plant stage under natural incidence of the yellow rust disease in field condition of Sari, Iran. Nine genotypes of 8252 and 8320, 8395, 8396, 8103, 8150, 8348, 8426 and 8472 reported resistant. The presence of *Yr1*, *Yr4*, *Yr10*, and *YrSP* genes were postulated for these genotypes. The presence of some unknown resistant genes suggested during comparison of resistant genotypes reaction and the virulence yellow rust differential set (Zahravi et al., 2019).

Then 106 Iranian wheat landraces were evaluated against yellow rust. New sources of seedling resistance were identified using 10 different *Pst* races including PstS10, PstS7, PstS2+V27, PstS11, PstS8, PstS3, PstS13, PstS6, PstS0. Sixteen genotypes (15%) mostly originated from the Northwest of Iran described resistance to all *Pst* isolates (Yazdani et al., 2020). Information on commercial wheat cultivars and their reaction to yellow rust is in Table 4.2. They commercialized from 2000 until now in Iran.

Table 4.2 Wheat cultivars released in Iran since 2000

No.	Cultivar	Year of release	Institute Origin	Growth habit	Resistance to yellow rust
1	Koohdasht	2000	DARI, CIMMYT	B, S	MR
2	Shahryar	2001	SPII, Iran	B, W	S
3	Azar2	2001	DARI, Iran	B, W	MS
4	Pishtaz	2002	SPII, Iran	B, S	R
5	Shiraz	2002	SPII, Iran	B, S	S
6	Tous	2002	SPII, IWWIP	B, F	S
7	Arya	2003	SPII, CIMMYT	D, S	R
8	Karkheh	2005	SPII,.ICARDA	D, S	R
9	Bam	2006	SPII, Iran	B, S	MR
10	Neishabour	2006	SPII, Iran	B, S	MS
11	Sistan	2006	SPII, Iran	B, S	MS
12	Arta	2006	SPII, Iran	B, S	MS
13	Moghan3	2006	SPII, Iran	B, S	S
14	Drya	2006	SPII, Iran	B, S	S
15	Bahar	2007	SPII, ICARDA	B, S	S
16	Dena	2007	SPII, CIMMYT	D, S	R
17	Rasad	2007	DARI, Iran	B, W	MR-MS
18	Pishgam	2008	SPII, Iran	B, F	R
19	Sivand	2009	SPII, Iran	B, S	R
20	Parsi	2009	SPII, Iran	B, S	R
21	Uroum	2009	SPII, Iran	B, W	R
22	Arg	2009	SPII, Iran	B, F	MS
23	Homa	2009	DARI, Iran	B, W	S
24	Morvarids	2009	SPII, CIMMYT	B, S	R
25	Behrang	2009	SPII, CIMMYT	D, S	R
26	Dehdasht	2009	DARI, Italy	D, S	MR
27	Saji	2009	DARI, ICARDA	D, S	R
28	Zare	2010	SPII, IWWIP	B, F	R
29	Mihan	2010	SPII, Iran	B, W	R
30	Aflak	2010	SPII, CIMMYT	B, S	R
31	Ohadi	2010	DARI, Iran	B, W	S
32	Rijaw	2011	DARI, IWWIP	B, F	R
33	Karim	2011	DARI, ICARDA	B, S	R

(continuous)

No.	Cultivar	Year of release	Institute Origin	Growth habit	Resistance to yellow rust
34	Ofog	2012	SPII, Iran	B, S	MR
35	Sirvan	2012	SPII, CIMMYT	B, S	R
36	Gonbad	2013	SPII, Iran	B, S	MR
37	Chamran2	2013	SPII, Iran	B, S	R
38	Takaab	2013	DARI, Iran	B, W	MR
39	Mehrgan	2014	SPII, Iran	B	R
40	Shoush	2014	SPII, Iran	B	MR
41	Narin	2014	SPII, Iran	B	MR-MS
42	Baharan	2014	SPII, Iran	B	R
43	Shabrang	2014	SPII, Iran	D	R
44	Baran	2014	DARI, Iran	B, W	MS
45	Ghaboos	2014	DARI, CIMMYT	B, S	R
46	Barat	2015	SPII, Iran	B	R
47	Heydari	2015	SPII	B	R
48	Rakhshan	2015	SPII	B	R
49	Shavver	2015	SPII	B	R
50	Hana	2015	SPII	D	R
51	Aftab	2015	DARI, Iran	B, S	R
52	Hashtrood	2015	DARI, Iran	B, W	MS
53	Saein	2015	DARI, Iran	B, W	R
54	Ehsan	2016	SPII	B	R
55	Khalil	2016	SPII	B	R
56	Sadra	2016	DARI, Iran	B, W	MS
57	Talaei	2017	SPII	B	R
58	Sarang	2017	SPII	B	MR
59	Zarineh	2017	SPII	B	MR
60	Tirgan	2017	SPII	B	R
61	Aran	2017	SPII	D	MR
62	Ivan	2017	DARI, Iran	B, F	R
63	Aseman	2017	DARI, Iran	B, S	MR-MS
64	Paraw	2017	DARI, Iran	B, W	MR
65	Zahab	2017	DARI, ICARDA	D, S	R
66	Saverz	2017	DARI, CIMMYT	D, S	MR

(continuous)

No.	Cultivar	Year of release	Institute Origin	Growth habit	Resistance to yellow rust
67	Kalateh	2018	SPII	B	R
68	Barzegar	2018	SPII	B	MR-MS
69	Torabi	2018	SPII	B	R
70	Meraj	2018	SPII	B	MR
71	Setareh	2018	SPII	B	MR
72	Kamal	2018	DARI, Iran	B, W	MSS
73	Shalan	2018	DARI, Iran	B, W	MSS
74	Rahmat	2018	DARI, CIMMYT	B, W	MSS
75	Varan	2018	DARI, Iran	B, W	MR-MS
76	Amin	2019	SPII	B	R
77	Farin	2019	SPII	B	R
78	Heyran	2019	SPII	B	MR
79	Sahar	2019	SPII	B	MR
80	Taban	2019	SPII	D	MR
81	Sana	2019	SPII	D	MR
82	Mehr	2019	DARI, Iran	B, W	R
83	Paya	2019	DARI, CIMMYT	B, S	MR
84	Kabir	2019	DARI, CIMMYT	B, S	MR

Note: Bread Wheat (B), Durum Wheat (D), Spring type (S), Winter type (W), Facultative type (F), Resistant (R), Moderately resistant (MR), Moderately susceptible (MS), Susceptible (S).

4.5 Pathogen Population Genetics and Physiological Races Identification of Yellow Rust

Study on the yellow rust physiologic race identification in Iran had started from 1963 with detection of two biotype of 20A and 25A (Niemann et al., 1968) . Macer (1972) noted that yellow rust was important in cooler parts of Yugoslavia, Egypt, Turkey and Iran (Macer, 1972). There were two high yielding popular cultivars during that time in Iran. Bamdadian reported 14/8, 14/8A, 19, 2D, 20A and 25A from all over Iran (Bamdadian, 1972). Later Bamdadian et al. (1991) reported the new race of wheat yellow rust that made Azadi and Ghods cultivars susceptible as 96E16. Then the investigation of Torabi et al. (2001) during seven-year study on *Pst* isolates collected from all over Iran revealed more virulence incidence for *Yr2, Yr6, Yr7, Yr9* and *YrA* and there was no virulence for *Yr1, Yr4, Yr5* and *Yr10* (Torabi et al., 2001). Four years later new race of *Pst* were identified which were virulent on Chamran, the popular cultivar of that time. Based on the isolates from Kermanshah and Fars provinces, these were new race of 166E134A+, reported for the first time from Iran. New races with virulence against the *Yr2*, *Yr2*+, *Yr7*, *Yr7*+, *Yr2.6*+, *Yr9*, *Yr2.9*+, *Yr24*, *YrSD*, *YrA* genes were confirmed (Afshari et al., 2004).

Two-year virulence monitoring of *Pst* during 2003 and 2004 denoted the Pathotypes 6E6A+, 6E22A+, 6E130A+, 6E134A+, 6E142A+, 6E158A+, 134E130A+ and 134E142A+ as more common despite identification of twenty-seven pathotypes in greenhouse tests (Afshari, 2008). Virulence on plant/s with gene/s *Yr2, Yr6, Yr7, Yr8, Yr9, Yr24, Yr25, YrSD, YrSP, Yr3N, Yr2+, Yr6+, Yr9+, Yr7+, Yr32+* and *YrA* was also reported under greenhouse conditions. Virulence on cultivars with *Yr2, Yr6, Yr7, Yr9, YrA* and *Yr24* genes were reported in high frequency. Cultivars with *Yr1, Yr3V, Yr4, Yr5, Yr10* and *YrSU* genes were reported resistant. Three years field study showed virulence on wheat genotypes Heines Kolben (with genes *Yr2* and *Yr6*), Kalyansona (*Yr2*), Lee (*Yr7*), Avocet R (*YrA*), Federation*4/Kavkaz (*Yr9*) and TP1295 (*Yr25*). In the trap nurseries, virulence was not reported on cultivars with *Yr1, Yr3V, Yr3N, Yr4, Yr5, Yr8, Yr10, Yr18, Yr24, Yr32+, YrSP, YrSD* and *YrSU* genes (Afshari, 2008). Another research on the virulence and molecular polymorphism of *Pst*, using a collection of 86 isolates from all around the Iran revealed the most frequent race as 178E0A- followed by race 64E241A+. No virulence on the commercial cultivars having *Yr3*, and *Yr5* genes was detected but virulence for *Yr2, Yr24, Yr7* and *YrA* was reported all around the country. AFLP test categorized the isolate in 19 groups, where 10 groups had only one member. Significant relationship between geographic origins and AFLP fingerprinting groups was reported. However there were not certain relationship between races and these groups (Rabaninasab et al., 2008). More than 65% genetic similarity between most of populations and West/Northwest population was reported. This similarity were described as effect of Sudanian and Mediterranean flows in transferring spores from Northwest to North and Southwest of Iran (Rabaninasab et al., 2008).

Samples collected from various environments during 2008 to 2009, revealed that out of 29 isolates the pathotypes 6E2A+, 6E6A+, 6E130A+ and 6E150A+ were identified. In 2008, the most and the least virulence potency was related to isolates Ahvaz2 and Gharakhil that could overcome 12 and 5 resistance genes of the host plant, respectively. In 2009, the most virulence potency related to isolates from BayekolaII, ToroghII, Borojerd that could overcome 10 and the least ZarghanII that was virulent on 4 resistance genes. Over all the majority of isolates with high frequency were reported virulent on plant/s with *Yr2, Yr2+, Yr7, Yr7*+ and *YrA genes*. No virulence was detected on plant/s with *Yr1, Yr3, Yr4, Yr5, Yr9+, Yr10, Yr15, YrCV, YrSP, YrSU* and *YrND genes* (Niazmand and Afshari, 2010).

During study of the *Pst* pathogenicity the race 166E254A + *Yr27*+ with 62.50% of pathogenesis was the most aggressive race from Torogh, Khorasan razavi province and the race 6E134A+ with 33.34% of pathogenesis was the least aggressive race from Zarghan, Fars province. Virulence of *Pst* isolates on all the wheat cultivars containing genes *Yr2, Yr6, Yr7, Yr9, Yr18, YrA* was observed but there was not virulent isolate on wheat cultivars with genes *Yr1, Yr4, Yr5, Yr10, Yr15, YrSU*. Also, pathogenesis on cultivars bearing *Yr3, Yr24, YrSP* (2.7%) was low but on cultivars with genes *Yr2, Yr6, Yr7, Yr9, Yr18, Yr4* (100%) or *Yr17* (97.3%) was reported high (Pornamazeh et al., 2013). Three races; 6E6A+, 6E150A+ and 198E150A+ of yellow rust were reported from Ardebil, Zarghan and Sari wheat cultivation (Omrani et al., 2013). Comprehensive monitoring of *Pst* races throughout the country reported by Afshari that from 2008 to 2010 the races 6E6A+, 6E10A+ and 6E0A+b was more common. Races 0E0A+ was less aggressive than races 166E158A+ and 134E158A+ with virulence on 11 known genes. During that time virulence on plant/s with gene/*sYr1, Yr2, Yr4, Yr6, Yr7, Yr8, Yr9, Yr10, Yr25, Yr27, YrSU, YrSD, YrND, Yr3, Yr2+, Yr6+, Yr9+, Yr7+, YrCV* and *YrA* was detected. More than 70% of isolates described with virulence on plant/s bearing *Yr2, Yr7, Yr9* and *YrA* genes. Virulence on plant/s with *Yr3, Yr5* and *YrSP* were not detected. Result of greenhouse test revealed frequency of less than 7% for virulence against *Yr1, Yr4, Yr10, YrCV (32+)* and *YrSD* genes. Virulence Frequency to other wheat genotypes reported between 8% to 100% (Afshari, 2013). During another research on race analysis of 37 isolates of *Pst* from most important wheat-growing areas of Iran, race 166E254A + *Yr27+* with 62.50% of pathogenesis was reported as the most aggressive race from Torogh (Mashhad), and the race 6E134A+ with 33.34% of pathogenesis was the weak race from Zarghan. Base on their results, all isolates were virulent against *Yr2, Yr6, Yr7, Yr9, Yr18, YrA*. The resistance of plants with genes *Yr1, Yr4, Yr5, Yr10, Yr15, YrSU* were effective against all isolates (Pornamazeh et al., 2013). Race analysis of *Pst* isolates collected over wheat field of Iran during 2012-2013 growing season revealed that all isolates had pathogenicity on the differential variety carrying gene *YrA* but pathogenicity was not reported on the differential varieties carrying genes *Yr5, Yr10, Y15*, and *YrSP*. According to observation, race 132E156A+, *Yr27* from Ahvaz, Khozestan province, southwest Iran was reported the most aggressive and race 2E2A+ from Mashhad, Khorasan province, northeast was the least aggressive race (Soweizy et al., 2016).

Provincial monitoring on two yellow rust trap nurseries in Zarghan and Mammassani during 2003-2013 has been conducted and among the differential cultivars, virulence was reported for *Yr2, Yr6, Yr7, Yr9, Yr17, Yr25, Yr26, Yr27*, *YrA* and *YrSU*. According to results of monitoring, the emphasis was placed on annual survey on responses of commercial wheat cultivars to yellow rust together with monitoring of pathogen virulence status in Fars province. This information can effectively aid on-time warning to replace susceptible cultivars and to use effective control measures to reduce the loss of the disease, and planning short and long time wheat breeding programs (Zakeri et al., 2017).The field base research at Allarough Agricultural Research Station in Ardabil, Iran revealed that yellow race/races had virulence for *Yr2, Yr6, Yr7, Yr9, Yr17, Yr21, Yr22, Yr23, Yr24, Yr25, Yr26, Yr27, Yr31, Yr32, Yr A* and *YrSU* resistance genes (Safavi, 2021).

4.6 Integrate Management of Wheat Yellow Rust

Although there are several ways for the management of wheat yellow rust but the use of resistant cultivars is the most effective, environmentally friendly, and economic way for disease control. Yellow rust has high virulence diversity and evolutionary aptitude. Migration of new races by cyclonic wind and mutation in a population of *Pst* in yellow rust hot spot geographical region besides selection pressure on pathogen could lead to the emergence of new races of yellow rust and increase in the proportion of virulent races and pathotype in the population of *Pst*. New virulent pathotypes with the new structures could overcome the resistance genes in commercial resistant cultivars, spread in the wheat fields, and cause disease epidemics. Therefore, integrated disease management strategy should adopt for yellow rust control.

Effective and sustainable control methods against yellow rust disease highly correlate to the knowledge of the disease in the country that was reviewed in previous sections.

Several recommendations for cultural practice confirmed in Iran. It commended to use adequate seed planting density for each variety with avoidance of excessive nitrogen fertilizer (Dadrezaei and Torabi, 2016). Some scientists warned for early planting and endorsed removing volunteer wheat and host weeds along with the wise use of nitrogen fertilizer.

In case of emergency conditions of disease epidemics, the use of agrochemicals from Triazole fungicides, which belong to the sterol demethylation inhibitor (DMI) group, could be adopted. Agrochemical spray could be considered till the early grain filling stage on cultivars with more than two tons per hectare yield estimation to be economically justifiable (Sadravi, 2014; Dadrezaei and Torabi, 2016).

Monitoring and practice forecasting for disease incidence and epidemic prediction are obligatory for integrated disease management of yellow rust in Iran. The main factors for the disease epidemic forecast are climatic factors such as temperature, humidity, intensity and direction of the wind, cultivars' response to disease and their growth stage, the existence of volunteer wheat in summer, planting time, and level of nutrition to determining the numbers of fungicide application. Observation of *Pst* on volunteer wheat in summer is a sign. After increasing the minimum temperature over 7℃ when the mean temperature during five days period becomes 13~15℃ and relative humidity of more than 70% and at least two days with more than five millimeters of rain, there is the probability of yellow rust epidemic during the following three to five days. Therefore, farmers that sowed susceptible or moderately susceptible cultivars need to be prepared for the agrochemical spray and start treatment when the *Pst* reddish to orange blister-like swellings called pustules are detected on 10%~20% of the leaves (Sadravi, 2014).

The following steps for integrated management (IPM) of yellow rust disease in Iran are highlighted:

- Establishment of IPM technical commission before the start of the growing season at the provincial level for integrated management and prevention of excessive agrochemical use.

- Distribution of certified resistant cultivars seeds among farmers according to climate-suitable type on time. Cultivars Morvarid, Ghonbad, and Shiroodi are suitable for warm and humid climates. Cultivars Pishgham, Urum, Zare, and Mihan are proper for cold climates. Then Cultivars Parsi, Bahar, Sivand, and Sirvan are appropriate for temperate climate, and cultivars Mehrghan, Chamran2, and Aflak are proper for the warm and dry climate of southern Iran.

- Effective yellow rust monitoring web and capacity building with the training personnel needs to form.

- Weekly monitoring of the wheat field and reporting the main factors for yellow rust disease forecast with emphasis on susceptible cultivars based on the *Pst* race identification results.

- Preparation of adequate agrochemicals for point application of agrochemicals such as the Tilt and Folicur that could be applied at one liter per hectare according to the result of forecasting model before grain filling stages.

- A single application of fertilizer as a top dressing, adequate irrigation, and avoiding the waterlogging on the field (Dadrezaei and Torabi, 2016).

4.7 Future Avenues to Work on

Screening of wheat genotypes for resistance and monitoring of *Pst* virulence diversity over multi-environments is very important for breeding future resistant wheat cultivar. Although there are more genes to choose from to develop cultivars for deployment against ever changing yellow rust pathogen, different gene combinations must be encouraged for different agroecological zones of wheat cultivation.

Precision breeding using molecular marker-assisted selection, despite expensive equipment and expertise requirement, accelerates the breeding process. Such strategies must be opted in case of wheat breeding. Attention must be paid to environmental concern and green agricultural production management strategies along with diversification of the resistance gene pool.

Identifying new genes and characterizing their effects could be a lengthy process, but it can be achieved through regional cooperation to benefit from the state-of-the-art facilities available there. The selection pressure for fungicide-resistant strains could be neutralized by diversification of safer agrichemicals development for the use in integrated management of yellow rust. Monitoring of the pathogen populations should continue, and effective bio-fungicides that can be used for organic farming and environmentally protected areas should be developed.

Developing suitable cultivars with resistance to yellow rust is necessary and environmentally friendly technique but need continuous monitoring of the pathogen and evaluating breeding material with broadening the genetic base of commercial varieties that need proper financing. Thus, regional and global collaborative efforts should be encouraged for this purpose.

References

Abbasi M, Hedjaroude GA, Scholler M, et al., 2005. Taxonomy of *Puccinia striiformis* sl in Iran. *Rostaniha*, 5(2):199-224.

Afshari F, 2013. Race Analysis of *Puccinia striiformis* f. sp. *tritici* in Iran. *Archives of Phytopathology and Plant Protection*, 46(15):1785-1796.

Afshari F, Torabi M, Malihipour A, 2004. Appearance of a New Race of *Puccinia striiformis* f. sp. *tritici* in Iran. *Seed and Plant*, 19(4):543-545.

Afshari F, 2008. Prevalent Pathotypes of *Puccinia striiformis* f. sp. *tritici* in Iran. *J. Agric. Sci. Technol.*, 10:67-78.

Afzal SN, Haque M, Ahmedani M, et al., 2007. Assessment of Yield Losses Caused by *Puccinia striiformis* triggering Yellow Rust in the Most Common Wheat Varieties. *Pakistan Journal of Botany*, 39(6): 2127-2134.

Ahmad S, Afzal M, Noorka IR, et al., 2010. Prediction of Yield Losses in Wheat (*Triticum Aestivum* L.) Caused by Yellow Rust in Relation to Epidemiological Factors in Faisalabad. *Pak. J. Bot.*, 42(1): 401-407.

Ajirloo T, Torabi M, Safavi S, 2016. Evaluation of Partial Resistance Components in Some Promising Wheat Lines of Cold Climate Zone to Yellow Rust Disease in Field Condition in Ardebil, Iran. *Seed and Plant Improvement Journal*, 32(3): 347-367.

Ali S, Gladieux P, Leconte M, et al., 2014. Origin, Migration Routes and Worldwide Population Genetic Structure of the Wheat Yellow Rust Pathogen *Puccinia striiformis* f. sp. *tritici*. *PLoS Pathogen*, 10 (1): e1003903. DOI:10.1371/journal.ppat.1003903.

Ali S, Sharma S, Leconte M, et al., 2018. Low Pathotype Diversity in a Recombinant *Puccinia striiformis* Population Through Convergent Selection at the Eastern Part of Himalayan Centre of Diversity (Nepal). *Plant Pathology*, 67:810-820.

Ali S, Rodriguez-Algaba J, Thach T, et al., 2017. Yellow Rust Epidemics Worldwide were Caused by Pathogen

Races from Divergent Genetic Lineages. *Frontiers in Plant Science*, 8:1058.

Arshad Y, Abbasi Moghadam A, Alitabar AR, et al., 2012. Evaluation of Resistance to Yellow Rust in Part of Bread Wheat Collection of National Plant Gene Bank of Iran. Paper presented at the The 13th international cereal rust and powdery mildews conference, Beijing, China.

Bakhtiar F, Farshadfar E, Sarbarzeh M, et al., 2016. Evaluation of Resistance to Yellow Rust in Doubled Haploid Lines of Bread Wheat. *Seed and Plant Improvement Journal*, 31(4): 679-698.

Bamdadian A, Rajabi S, Rahbari B, 1991a. New Virulence of Yellow Rust of Wheat (*Puccinia striiformis*) in Iran. Paper presented at the Proceedings of the 10th Plant Protection Congress of Iran 1-5 Sep. 1991, Kerman, Iran.

Bamdadian A, Rajaei S, Rahbari B, 1991b. New biotype of whear yellow rust in Iran. Paper presented at the 10th Iranian Plant Protection Congress, University of Shahid Bahonar, Kerman, Iran.

Bamdadian A, Torabi M, 1978. Epidemiology of Wheat Stem Rust in Southern Areas of Iran in 1976. *Iranian Journal of Plant Pathology*, 14(1/4), 20-19.

Bamdadian A, 1972. Physiologic Races of *Puccinia striiformis* West. Paper presented at the 6th European and Mediterranean Cereal Rust Conference, Parha-Czecholvakia, Iran.

Biffen RH, 1905. Mendel's Laws of Inheritance and Wheat Breeding. *The Journal of Agricultural Science*, 1(1): 4-48.

Bimb H, Johnson R, 1997. Breeding Resistance to Yellow Rust in Wheat. *Wheat Special Report No.41*. Mexico: CIMMYT.

Chen X, 2005. Epidemiology and Control of Yellow Rust [*Puccinia striiformis* f. sp. *tritici*] on Wheat. *Canadian Journal of Plant Pathology*, 27(3):314-337.

Chen X, 2020. Pathogens Which Threaten Food Security: *Puccinia striiformis*, the Wheat Yellow Rust Pathogen. *Food Security*, 4: 1-13.

Dadrezaei S, Jafarnezhad A, Lakzadeh I, et al., 2018. Evaluation of Tolerance to Yellow Rust Disease in Some Selected Bread Wheat cultivars. *Seed and Plant breeding Journal*, 34(1):125-142.

Dadrezaei S, Nazari K, 2015. Detection of Wheat Rust Resistance Genes in Some Iranian Wheat Genotypes by Molecular Markers. *Seed and Plant Improvement Journal*, 31(1):163-187.

Dadrezaei T, Torabi M, 2016. Management of Wheat Rusts. *Plant Pathology Science*, 5(2): 81-89.

Dehghani H, Moghaddam M, 2004. Genetic Analysis of the Latent Period of Yellow Rust in Wheat Seedlings. *Journal of Phytopathology*, 152(6):325-330.

Dehghani H, Torabi M, Moghadam M, et al., 2005. Biplot Analysis of Diallel Cross Data for Infection Type of Wheat Yellow Rust. *Seed and Plant*, 21(1):123-138.

Dubcovsky J, Dvorak J, 2007. Genome Plasticity a Key Factor in the Success of Polyploid Wheat under Domestication. *Science*, 316(5833):1862-1866.

El Baidouri M, Murat F, Veyssiere M, et al., 2017. Reconciling the Evolutionary Origin of Bread Wheat (*Triticum aestivum*). *New Phytologist*, 213(3):1477-1486.

Elahinia S, Tewari J, 2005. Assessment of Two Different Sources of Durable Resistance and Susceptible Cultivar of Wheat to Yellow Rust (*Puccinia striiformis* f. sp. *tritici*). *Caspian Journal of Environmental Sciences*, 3(2):117-122.

Elahinia SA, 2000. Assessment of Urediniospore Germination of *Puccinia striiformis* at Various Temperatures on Agar and Detached Leaves of Wheat. *Journal of Agricultural Science and Technology (JAST)*, 2(1): 41-47.

Esfandiari E, 1947. Cereal Rusts in Iran. *Entomology and Phytopathology Journal*, 4: 67-76.

Ghaffari A, JalalKamali MR, 2013. Wheat Productivity in Islamic Republic of Iran: Constraints and Opportunities. In Paroda R, Dasgupta S, Mal B, Singh SS, Jat ML, Singh G (Eds.), *Improving Wheat Productivity in Asia*. Thailand: APARRI and FAO.

Jin Y, Szabo LJ, Carson M, 2010. Century-old Mystery of *Puccinia striiformis* Life History Solved with the Identification of *Berberis* as an Alternate Host. *Phytopathology*, 100(5): 432-435.

Khanfri S, Boulif M, Lahlali R, 2018. Yellow Rust (*Puccinia striiformis*): a Serious Threat to Wheat Production Worldwide. *Notulae Scientia Biologicae*, 10(3):410-423.

Khazra H, Bamdadian A, 1974. The Wheat Disease Situation in Iran. Paper presented at the Fourth FAO/ Rockefeller Foundation Wheat Seminar, Tehran, Iran.

Khodarahmi M, Mohammadi S, Bihamta M, et al., 2014. Inheritance and Combining Ability of Yellow Rust Resistance in Some Bread Wheat Commercial Cultivars and Advanced Lines. *Seed and Plant Improvement Journal*, 30(3):531-544.

Macer R, 1972. The Resistance of Cereals to Yellow Rust and its Exploitation by Plant Breeding. *Proceedings of the Royal Society of London. Series B. Biological Sciences*, 181(1064): 281-301.

Malihipour A, Torabi M, 2008. Reaction of Rainfed Wheat Advanced Lines and Cultivars at their Seedling Stages to Two Races of *Puccinia striiformis* f. sp. *tritici*. *Iranian Journal of Field Crop Science (Iranian Journal of Agricultural Sciences),* 39(1): 193-202.

Mehdinia F, Alaei H, Sedaghati E, et al., 2016. Distribution and Genetic Diversity of Aecial Infection on Barberry and its Importance to Wheat Yellow Rust Disease in Lorestan Province. *Iranian Journal of Plant Pathology*, 52(2).

Nazari K, Torabi M, Dehghan M, et al., 2001. Pathogenicity of *Puccinia striiformis*, and Reactions of Improved Cultivars and Advanced Lines of Wheat to Yellow Rust in Northern Provinces of Iran. *Seed and Plant*, 16(4): 393-424.

Nazari K, Torabi M, Saidi A, et al., 2001. Seedling and Adult-Plant Resistance to Yellow Rust Wheat in a Global Environment. Berlin: Springer.

Niazmand AR, Afshari F, 2010. Study on Pathotype Diversity and Virulence Factors of *Puccinia striiformis* f. sp. *tritici*, the Causal Agent of Wheat Yellow Rust in Iran. *Journal of Microbial World*, 3(1): 63-73.

Niemann E, Scharif G, Bamdadian A, 1968. Die Getreideroste in Iran. *Wirtsbereich, untercheidung Bedeutung*

Bekampfung. *Entomologie et Phythopathologie Appliquees*, 27: 25-41.

Omrani A, Khodarahmi M, Afshari F, 2013. Evaluation of Resistance to Yellow Rust in Some Wheat Advanced Lines. *Seed and Plant Improvement Journal*, 29-1(4): 761-776.

Parlevliet J, 1985. Resistance of the Non-race-specific Type. In Roelfs A P and Bushnell WR (Eds.), *The cereal rusts. Diseases, Distribution, Epidemiology, and Control* (Vol. II). Amsterdam: Elsevier.

Pornamazeh P, Afshari F, Khodarahmi M, 2013. The Genetic of Pathogenicity of *Puccinia striiformis* f. sp. *tritici* the Cause's Agent of Wheat Yellow Rust Disease in Iran. *Archives of Phytopathology and Plant Protection*, 46(12): 1497-1507.

Rabaninasab H, Okhovat M, Torabi M, et al., 2008. Virulence and Molecular Diversity in *Puccinia striiformis* f. sp. *tritici* from Iran. *Journal of Plant Protection*, 22(2): 47-60.

Roelfs A, Singh R, Saari E, 1992. Rust Diseases of Wheat: Concepts and Methods of Disease Management (CIMMYT).

Sadravi M, 2014. Forecasting Model of Wheat Yellow Rust. *Plant Pathology Science*, 3(1): 62-74.

Safavi SA, 2015. Effects of Yellow Rust on Yield of Race-specific and Slow Rusting Resistant Wheat Genotypes. *Journal of Crop Protection*, 4(3): 395-408.

Safavi SA, Torabi M, Afshari F, 2008. Resistance Reaction of Promising Wheat Genotypes of Cold Zone to Yellow Rust. *Journal of Research in Agricultural Science*, 4(1): 93-103.

Sinha P, Chen X, 2021. Potential Infection Risks of the Wheat Yellow Rust and Stem Rust Pathogens on Barberry in Asia and Southeastern Europe. *Plants* (Basel), 10(5):957. DOI: 10.3390/plants10050957.

Solh M, Nazari K, Tadesse W, et al., 2012. The Growing Threat of Yellow Rust Worldwide. Paper presented at the Borlaug Global Rust Initiative (BGRI) conference, Beijing, China.

Soweizy M, Afshari F, Rezaee S, 2016. The Pathogenicity of *Puccinia striiformis* f. sp. *tritici* in Iran in 2012-2013 Growing Season. *Plant Protection (Scientific Journal of Agriculture)*, 39(2): 13-22.

Torabi M, 1980. Factors Affecting an Epidemic of Yellow Rust on Wheat in the North-western and Western Regions of Iran. Paper presented at the Fifth European and Mediterranean Cereal Rusts Conference, Bari and Rome, Italy.

Torabi M, 2004. Epidemiology of Wheat Yellow Rust (*Puccinia striiformis* f. sp. *tritici*) in Iran. Paper presented at the Meeting the Challenge of Yellow Rust in Cereal Crops Islamabad, Pakistan.

Torabi M, Mardoukhi V, Nazari K, et al., 1995. Effectiveness of Wheat Yellow Rust Resistance Genes in Different Parts of Iran. *Cereal Rusts and Powdery Mildews Bulletin*, 23(1): 9-12.

Torabi M, Nazari K, 1997. Seedling and Adult Plant Resistance to Yellow Rust in Iranian Bread Wheats. Paper presented at the Wheat: Prospects for Global Improvement, Ankara, Turkey.

Torabi M, Nazari K, Afshari F, et al., 2001. Seven Years Pathotype Survey of *Puccinia striiformis* f. sp. *tritici* in Iran. Paper presented at the First Regional Yellow Rust Conference for WANA. Karaj, Iran. ICARDA.

Venske E, dos Santos RS, Busanello C, et al., 2019. Bread Wheat: a Role Model for Plant Domestication and

Breeding. *Hereditas*, 156(1), 1-11.

Viani A, 2009. Prevalence of Yellow Rust Disease in Irrigated Wheat Fields in the Cold Regions of Kohghilooye and Boyerahmad Province. *Journal of Agricultural Sciences and Natural Resources*, 16 (Special Issue 2):190-196.

Wellings CR, 2011. Global Status of Yellow Rust: a Review of Historical and Current Threats. *Euphytica*, 179(1): 129-141.

Yazdani M, Patpour M, Yassie M, et al., 2020. Evaluation of Resistance to Yellow and Stem Rust in some Native Iranian landraces of wheat. Paper presented at the BGRI workshop online.

Zahravi M, Afshari F, 2018. Identification of Resistance Sources to Yellow Rust in Some Genotypes of Bread Wheat Collection of the National Plant Gene Bank of Iran. *Seed and Plant Breeding*, 34-1(1): 1-14.

Zahravi M, Afshari F, Ebrahimnejad S, 2019. Study of Genetic Diversity of Resistance to Yellow Rust in Bread Wheat Germplasm. *Modern Genetics Journal* , 14(3): 263-274.

Zahravi M, Asgharzadeh P, Afshari F, et al. Study of Relationships among Components of Resistance to Yellow Rust (*Puccinia striiformis* f. sp. *tritici*) in Iranian Wheat Landraces. *Modern Genetics Journal*, 4(4): 33-43.

Zahravi M, Bihamta MR, 2010. Estimation of Gene Effects and Combining Ability of Latent Period of Yellow Rust in Advanced Lines of Wheat. *Iranian Journal of Genetics and Plant Breeding*, 1(1): 52-58.

Zahravi M, Ebrahimnejad S, Afshari F, 2012. Evaluation of Field Based Partial Resistance and Relationship between Resistance Components of Bread Wheat Germplasm to Yellow Rust. *Seed and Plant Improvement Journal*, 28(4): 663-684.

Zahravi M, Taleei A, Zeynali H, et al., 2004. Diallel Analysis of Infection Types of Two Yellow Rust Pathotypes, 6E130A+ and 166E42A+, in Some Advanced Lines of Wheat. *Seed and Plant*, 20(1): 73-88.

Zakeri A, Yassaie M, Afshari F, et al., 2017. Surveying Virulence of the Causal Agent of Wheat Yellow Rust (*Puccinia striiformis* f. sp. *tritici*) and Determining Reaction of Commercial Wheat Cultivars Over the Past Decade in Fars, Iran. *Iranian Journal of Plant Pathology*, 52(3): 297-316.

Zhao J, Wang M, Chen X, et al., 2016. Role of Alternate Hosts in Epidemiology and Pathogen Variation of Cereal Rusts. *Annual Review of Phytopathology*, 54: 207-228.

Chapter 5
Wheat Yellow Rust in Iraq – Current Status and Future Challenges

Emad Mahmood Ghaleb Al-Maaroof

Biotechnology and Crop Science Dept., College of Agricultural Engineering Sciences, University of Sulaimani, IKR, Iraq

Email: emad.ghalib@univsul.edu.iq

Abstract: Wheat yellow (stripe) rust disease incited by the basidiomycetes fungus *Puccinia striiformis* f. sp. *tritici* (*Pst*) is currently considered as one of the most destructive foliar diseases of wheat in many wheat-growing areas in Iraq. Yield losses may reach to more than 60% on the susceptible wheat cultivars. During the last five decades' extensive scientific research was conducted on the disease by the national scientists. The current chapter summarizes an overview on the available knowledges and information about the economic importance and distribution of yellow rust disease in Iraq, ranging from epidemiology to virulence spectrum, physiological races and the control measures. The information should be useful for devising strategies to eliminate the impact of the disease on wheat production, as well as understand the future risks of invasions at regional and global scale.

Keywords: Wheat Yellow (Stripe) Rust; *Puccinia striiformis* f. sp. *tritici* (*Pst*); Cereals Diseases; Disease Resistance; Iraq

5.1 Main Agro-ecological Features of Iraq

Wheat (*Triticum aestivum* L.) is the most important staple food of about 36% of the world population. It provides more than 20% of the food calories and protein consumed globally (Hawkesford et al., 2013). Wheat is one of the most important cereal crops in Iraq. Historically, wheat cultivation has originated in the north east of Iraq in an old historical village called "Charmo", dating back to about 7, 000 BC. Iraq is, thus, considered as a center of origin of wheat. Others refer to the northwest Iran or northeast Turkey as a result of a hybridization of tetraploid wheat *Triticum turgidum* (AB genome) and diploid *Aegilops tauschii*, the donor of the D genome (Bernardo, 2002). Wheat production areas in Iraq are divided into rain-fed and irrigated regions. Rain-fed areas are mainly located in the north including the Kurdistan region. Wheat output is higher in the irrigated regions, but the rain-fed regions account to a lesser extent for wheat production. An estimated 4.0 million tons of wheat were harvested in 2018, with a decrease of 14% from 2017, primarily due to adverse environmental conditions including biotic impact. Nearly the same amount of wheat had been purchased in 2018 (FAO, 2016; FAOSTAT, 2018). The Iraqi ministry of agriculture announced that Iraq had achieved self-sufficiency in wheat by producing more than 6 million tons of wheat in 2020. This was the result of cultivation of 2.25 million hectare of wheat lands including 1.5 million hectares of irrigated areas and 0.75 million hectares of rainfed areas (MOA, 2020).

Iraq occupies an area of approximately 438.317 km^2 with a current population of 40.53 m within the Middle East, Between latitudes 29 ° 5 ′ to 37 ° 15 ′ N and longitudes 38 ° 45 ′ to 48 ° 45 ′ E at the northern most point of Arabian gulf north of Saudi Arabia, west of Iran, east of Syria and south of Turkey. The altitude of the nation ranges from sea level in the south to as high as 3,500 m above sea level in the north eastern highlands. The land can be divided in to four main zones. The desert in the west and south west of Euphrates River, dominated by wide, flat sandy expanses. The uplands regions (Al-Jezirah) cover most of northern part of Iraq, ranging from altitude 100 ~ 450 m above sea level. The area is distinguished by deep valleys of the river. Irrigation in this region is more difficult than it is in the lower plain, and winter cultivation without irrigation is typically productive in this sector. The highlands in the north occupy much of the area in the north-east. Serious rises, interspersed with steppes, gives way to mountains as high as 4,000 m near the Iranian and Turkish borders. The fourth area is the alluvial plain which stretches from Baghdad in the north to the Arabian Gulf in the south, along the rivers Tigris and Euphrates. The whole region is a wide delta intersected by two river channels and irrigation canals including lakes and marshlands. The Mesopotamia plain, the ancient land of twin rivers, between Baghdad and Babylon is 30 ~ 40 m above sea level. Approximately 13% of Iraq's land area is listed as arable. Iraq's arable land has been rich and productive, particularly in the lower alluvial plain. Substantial amounts of arable land in the northwest upland area need irrigation. With its river systems, Iraq has the most plentiful water supplies in the region (Jaradat, 2002). Wheat and Barely account for almost half of the cultivated lands, Wheat is grown in 2,400 thousand hectares, *Triticum* spp: Bread wheat (*T. aestivum*), Durum wheat (*T. durum*) and Spelt (*T. spelta*) are well known in Iraq. Bread and durum wheat are widely grown and used for human food, among bread wheat, the main varieties are spring type and some are winter particularly in the high elevation lands, hard and soft, and red and white.

The climate in Iraq is mainly of the continental, subtropical semiarid type, with the north and north-eastern mountainous regions having a Mediterranean climate, with mild winters and dry, hot summers. The coldest month of the year is January, during which the average daily temperature in the plains varies from 2 ~ 15℃. The hottest months are in July and August during which the average daily temperature in the plains varies from about 24 ~ 43℃. The north-eastern uplands have cold winters with often heavy snowfalls. The average winter temperature in the western desert and the north-eastern foothills varies from 0 ~ 15℃ and the average summer temperature ranges from 22 ~ 38℃. The alluvial plain winter temperatures vary from 4 ~ 17 ℃ and summer temperatures vary from 29 ~ 43 ℃. Wheat production take place in a variety of diverse ecological conditions and mega environments, ranging from small farms in the highlands to large-scale fields in the lowlands (Fig 5.1). Spring wheat grown during the winter crop season from October/November to May/June in most wheat growing regions in the middle and south, while seed of spring and facultative winter wheat types grown during the winter crop season from November/December to June/July in high altitude regions. The lack of rainfall and intense heat makes a lot of Iraq desert. Due to extremely high evaporation, the soil and plants easily lose the little moisture from the rain, and the vegetation could not thrive without intensive irrigation. Intense cyclonic activity in the atmosphere is normally responsible for rainfall in Iraq. Rainfall occurs during the winter months, from December to February in most parts of the country and November to April in the mountains. Approximately 90% of the rainfall falls between November and April. The mean annual precipitation in the northern highlands varies from 850 ~ 950 mm and can reach 1,250 mm in the northeast highlands. The mean annual precipitation falls to 320 ~ 570 mm in the upland regions and to 100 ~170 mm in the western, southern and southwest regions.

Fig. 5.1 Wheat cultivation under diverse agro-ecological zones across Iraq

5.2 Wheat Diseases - The Main Biotic Constraints of Wheat Production in Iraq

Wheat crop is subjected to many important biotic constraints in Iraq. Among these constraints, diseases have been a major limiting factor for wheat production throughout history. Fungal diseases are the most important biotic constraints of wheat production in Iraq. Several diseases particularly rusts, smuts and septorial leaf blotches have drastically decreased grain yield and quality of wheat (Al-Adami, 1953; Al-Baldawi, 1993). The marketing value of the infected grains is also decreased (Al-Maaroof et al., 2005).

Rust diseases are the most widely spread of all wheat diseases in Iraq. Yellow, leaf and stem rust diseases caused by the fungi *Puccinia striiformis* f. sp *tritici* (*Pst*), *P. triticina* and *P. graminis* f. sp. *tritici* respectively are the most critical grain yield-limiting factor of wheat in Iraq. Yield losses due to the three rusts on wheat cultivars range from 10% ~ 70% depending on the resistance levels, the onset stage of the crop, the pathogen races and the environmental conditions (Al-Maaroof et al., 2001; Al-Azawi, 2005; Al-Maaroof and Nori, 2018). Among the three kinds of rusts, leaf rust is more important in the temperate zones and rapidly develop in the central parts of Iraq between 10 ~ 30 ℃ . The occurrence and distribution of leaf rust is more regular and uniform than other rust diseases in wheat is grown areas in the middle zones (Al-Maaroof et al., 1995, 2000). Leaf rust epidemics have been frequent in all the seasons except during 1997 and 2000 due to the dry conditions (Al-Maaroof et al., 2002). Leaf rust causes yield reductions up to 44% in commercial fields. Disease symptoms appear in the wheat fields from mid-April to a few times a little later. The symptoms are rarely seen in March (Al-Maaroof, 1997; Ali, 2006).

It is believed that the disease established in the wheat fields in the south and middle lands of the country and then moves to the north. Urediniospores are likely to comes from Saudi Arabia, Iran, Turkey and Syria. The alternate host, *Thialictrum* spp. have been found in northern parts of Iraq. But probably is not the primary source of inoculums (Al-Maaroof et al., 2002).

Yellow rust of wheat is the most prevalent rust disease in the cooler and humid wheat growing area. Formerly, yellow rust distribution was restricted in the northern parts of Iraq, particularly in the mountainous areas. This has rarely been seen in central and southern areas (Al-Baldawi et al., 1993; Al-Maaroof et al., 2001). In 1988 the disease was observed for the first time in the central area. Recently, several epidemics of the disease have occurred in all wheat growing areas particularly in irrigated fields. This was probably due to the appearance of new race(s) or to certain environmental factors that favored the development and incidence of the disease in the regions (Al-Baldawi, 1993; Al-Hamdani et al. , 2002; Al-Maaroof et al., 2003).

Stem rust is favored by humid and warmer weather 15 ~ 35 ℃ , usually distributed in the central and southern zones of Iraq. The importance of the disease is less than other rusts due to its appearance late in the season. The role of the alternate host "Barberry" is not recognized in Iraq (Al-Baldawi, 1983; Al-Maaroof, 2017).

5.3 Yellow Rust Incidence and Economic Importance

Yellow rust of wheat is the most widespread rust disease in the cooler and humid wheat growing area (Roelfs et al., 1992). Disease losses has been estimated to be at least 5.5 million tons per year at worldwide level (Beddow et al., 2015). The disease is one of the most important biotic constraints to sustainable wheat production in many wheat growing area's in Iraq. Distribution of yellow rust was formerly limited in the northern parts of Iraq. It has been rarely seen in the central and southern areas (Al-Baldawi et al., 1981; Al-Maaroof et al., 2001). The disease was well distributed in all wheat- growing areas of northern Iraq. The severity of the disease varies from season to season, depending on many factors, in particular wheat cultivars, the time of disease onset and the environmental conditions, including the amount of rainfall, humidity and temperature, and the amount of azotic fertilization used by farmers (Al-Beldawi, 1993).

Yellow rust was observed for the first time in some wheat fields in the middle zones in 1988. Later-on it has rapidly spread on the susceptible cultivars in this area and then moved to wheat growing areas in the south (Fig. 5.2). This was probably due to the appearance of new race (s) or to some environmental conditions which favored disease development and occurrence in the regions (Al-Maaroof et al., 2003). The recent accelerating climatic changes has already affected the complex biological interaction, which influenced the frequency and severity of disease epidemics and some studies indicated that climatic changes modifies disease and pest risks and increases uncertainly in risk predictions associated with climatic changes particularly in the wheat-*Pst* biological system (Chakraborty et al., 2010; Newton et al., 2010).

Yellow rust is considered to be the most prevalent and damaging rust annually in the region particularly in the Iraqi boarder countries. Torabi et al. (1995) reported that the disease was more serious in Iran in 1994 and caused more than 1.5 million tones damage in wheat field. El-Naimi and Mamluk (1995) stated that the early appearance of infection coupled with long wet spring mostly leads to high spread of yellow rust epidemics, which may cause high yield losses, yield reduction reached to approximately 108,000 tons in 1988 in Syria. While Kharouf (2009) referred to the excessive incidence of yellow rust in Syria in the years between 1990 and 2007 which caused significant losses in wheat production. The disease could affect 80% of wheat production area in Turkey and Syria especially in the cooler and humid regions (Karakas et al., 2009; Al-

Chaabi and Abu-Fadel, 2012). The high epidemics of the disease in 2010 resulted from high yield losses of 25MT in Iran (Kumarse, 2011). Al-Maaroof (1997) recorded 51% reduction in the grain yield of wheat due to yellow rust infection in Iraq in 1995, while grain yield loss was more than 40% in the susceptible wheat cultivars in 1996 (Al-Maaroof et al., 2001). Meanwhile yield reduction was more than 33% on cv. Maxipak under natural epidemic of the disease in 1998. The reduction was due to significant effects of the disease on the thousand grain weight (Al-Maaroof et al., 2003). Serious epidemics of yellow rust have been detected in most Iraqi wheat growing areas, particularly in the north in 2010, resulting in a high reduction in national grain yield. Yellow rust was quite destructive and was detected in early stages of wheat developments in the irrigated fields of the Iraqi–Turkish–Syrian Triangle in Duhok. Approximately ten thousand hectare of wheat fields were under risk in early stages of wheat development in this area. The disease severity reached up to 80MSS at the beginning of heading stage in cv. Tamuz 2 in a 250-hectare area, followed by 45S in a 100-hectare wheat field grown with Sham 6 at stem elongation stage (Al-Maaroof et al., 2012). Severe yellow rust epidemics resulted from 100% loss of grain yield on the susceptible wheat cultivar cv. SaberBeg at Bakrajo, Sulaimani in 2010 (Al-Maaroof et al., 2011).

Al-Bajalan (2012) recognized that yellow rust disease significantly decreased grain yield in wheat genotypes with diverse levels of resistance and susceptibility by 3.5%~35% under natural epidemic of the disease. Yield losses were mainly attributed to the significant effect of the disease in reduction of grain weight by 1.6%~17.7% and number of grains per spike by 2.3%~13%. Positive correlation between the AUDPC and the amount of grain yield losses were detected in wheat genotypes at high correlation level (R^2=0.9469). Furthermore, the disease significantly affected qualitative characters of wheat grains by decreasing protein and gluten contents (Al-Bajalan,2012; Al-Maaroof and Nori, 2019). Al-Mashhadani (2014) mentioned that various levels of disease incidence and severities of yellow rust were detected in most wheat fields during 2013 and 2014 growing seasons.

Yellow rust disease is currently considered to be the most important biotic constraint to sustainable wheat production in Iraq. Yield losses of 10% ~ 60% in Iraq are common, if the susceptible wheat cultivars are grown (Al-Maaroof et al., 2015). This is due to the ability of *Pst* to evolve rapidly into new race(s) and to migrate long distances by air-borne dispersal (Roelfs and Bushnell, 1985).

5.4 Epidemiology of Yellow Rust

The yellow rust epidemic has a very long history worldwide and has occurred in every place where the wheat cultivars are grown, each rust has its particular climatic adaptation and there is some evidence of change in that respect. yellow rust is principally a disease of wheat grown in cooler climates (2 ~ 15 ℃) which are generally associated with higher elevation, northern latitude or cooler years, however, the ability of the pathogen to survive in hot conditions became well known because of adaptation resulted from genetic variation (Roelfes et al., 1992). The three most important weather factors affecting epidemics of the disease are moisture, temperature and wind (Chen, 2005). In Iraq, yellow rust distribution was formerly restricted only in northern parts while recently the disease appears in most of wheat fields in the middle and southern parts of Iraq.

Severe yellow rust epidemics occurred in most Iraqi wheat growing fields during the 1998 growing season, leading to a significant decline in national grain production. The severity of the disease reached up to 93% on the susceptible cultivar Maxipak at an early stage of wheat development with a high infection rate of 0.358 per unit per day. The estimated loss of grain yield was more than 33% under natural epidemic conditions in the middle zones of Iraq, mainly due to the significant effect of the disease on grain weight (Al-Maaroof et al., 2001; Al-Maaroof et al., 2003). This was due to development of new virulence against *Yr9* gene which was predominant in most of wheat cultivars as a single major gene. Breakdown of *Yr9* resistance gene resulted in widespread epidemics in the Middle East and the Indian Subcontinent which caused considerable crop damage in the 1990s (Singh et al., 2004). The origin of this series of pandemics is considered to be the acquisition of virulence by the pathogen against *Yr9* arising in the red sea region as early as 1987 (Lowers et al., 1992), and might have established latter on in the region.

Severe outbreak of yellow rust disease has been observed in most of Iraqi's wheat-growing regions during the 2010 growing season, particularly in the northern and central regions as shown in Fig.5.3. Yellow rust epidemics differed depending on locations, growing wheat cultivars, stage of growth, onset of disease, irrigation systems and favorable environmental conditions, particularly cold nights and full or supplementary irrigations. The highest severity and type of infection (75HS) was recorded on cv. SaberBeg at stem elongation stage in Bakrajo, Sulaimania, while the incidence of disease reached up to 80% in the field at the same stage, followed by 80S on sham 6 in both Tanjaro and Halabja Taza with 90% and 80% incidence in each location respectively (Al-Maaroof et al., 2011). Kumarse (2011) reported that yellow rust outbreak was similar to that one which was already reported in 2010 in Syria, Iran and Turkey. The recent accelerating climatic changes have already affected the complex biological interaction, climate influence the frequency and severity of disease epidemics and some studies indicates that climatic changes modify disease and pest risks and increase uncertainty in risk predictions associated with climatic changes particularly in the wheat-*Pst* biological system. Yellow rust development on the commercial susceptible cultivars was high and resulted from severe epidemics in wheat fields. The highest infection rate reached to 0.337 on SaberBeg and caused 100% loss followed by 0.187 on Aras in the same location. About 10,000 hectare of wheat fields were under risk at early stages of wheat development in Duhok. Disease severity reached up to 80 MSS at the beginning of heading stage on Tamuz 2 in a 250-hectare area, followed by 45 S in a 100-hectare wheat field grown with Sham 6 at stem elongation stage. The yellow rust onset started early in Makhmore, Hawler. The severity of the disease reached up to 90S for Sham 6 in 80 Donum wheat fields at anthesis (Fig. 5.3). High disease severity and incidence were also reported early in the season for Rezgari, Sali, Azadi, Maxipak and sham 6 at different wheat fields. While the durum wheat cultivars Simetto and Fadda have been exploring resistant reactions at the adult plant stage in different wheat fields. High disease severity and incidence have been detected in various wheat cultivars in some fields in Kirkuk, Garmian, Diyala, Babel, Dheqar, Mesan, Wasit and Baghdad. The high rate of infection and disease incidence at the beginning of the season was due to the favorable environmental conditions, including high precipitation at the onset of the disease, which resulted in high relative humidity favorable for development of yellow rust disease on wheat. Furthermore, the dominant mean temperature was below 20℃ during the epidemic period in this area which is also more suitable for development of the disease (Rapilley, 1997; Al-Maaroof et al., 2012).

Yellow rust disease was known for a long time as micro cyclic rust, the alternate host of the disease had been unknown until 2009, when a team of scientists at the USDA-ARS Cereal Disease Lab led by Dr. Yue Jin confirmed that barberry (*Berberis* spp.) is an alternate host of yellow rust disease (Jin et al., 2010). Barberry was known as an alternate host of stem rust disease incited by *P. graminis* f. sp. *tritici*, and for many years, but their work helped to solve the "century-old mystery" of plant pathology (Jin et al., 2010). At the same time, recombination was discovered in the extended Himalayan population along with their high sexual reproduction capacity (Ali et al., 2010), suggesting a potential role in the Himalayan region. However, the presence of barberry in Iraq is not well described, while no evidence has been shown so far for the sexual reproduction (Bahri et al., 2009; Ali et al., 2014; Thach et al., 2016). That's why we believe that urediniospores of *Pst* enter Iraq from other neighboring countries such as Turkey, Iran, Syria, and other, countries by wind trajectories, as it is able to fly a long distance due to its air borne characters and resistance to short wave rays and unfavorable conditions (Roelfs et al., 1992; Chen, 2005). Over-summering of the pathogen can also occur by survival of the pathogen as a dormant mycelium or active sporulating pustules on volunteers or high altitude cultivated plants in the high mountain range (Al-Maaroof, 1997). Al-Maaroof and Noori (2018) observed inoculums of *Pst* arrives at the Sulaimani wheat fields in mid-March 2011. Approximately 1.75 urediniospore's were captured in one centimeter square during 24 hours in horizontal traps, but neither the spores were germinated on wheat leaves and nor the primary infection developed in March due to unfavorable environmental conditions to urediniospore germination and infection process. The first infection on the susceptible cultivars appeared on April 24. The number of urediniospores found on the trap has steadily increased with the development of the disease. The highest number of urediniospores reached to 67.5 spores/cm^2 on the horizontal stands when the disease severity on SaberBeg reached to 100% at milky stage on May 23, 2011. The number of urediniospores decreased with an increase in disease infection and plant maturity due to the decrease in relative humidity and precipitation, combined with a rise in temperature by the end of April. Number of urediniospores decreased with increase of the disease and plant maturity due to decrease in the relative humidity and the amount of precipitation coupled with increase in temperature by the end of April. Decreasing the RH to 50% and increasing the temperature over 25℃ led to decrease in urediniospore number, which produced on the susceptible tissues (Rapilly, 1979). Pustules continue producing urediniospores to one week and the highest production happened in the fourth day at a moderate temperature (Tomerlin et al., 1984).

The highest number of urediniospore may not be linked with increase in the infection rate because some of the urediniospores may be unviable or physiologically incompatible with the host plants. They also might have affected by the change of daily temperature and relative humidity since germination of urediniospores and invasion of host tissues needs 100% humidity for three to four hours (Tomerlin et al., 1984). Iraqi Kurdistan region is lying in the epidemiological zone 7, which is the potential source for the spread of new races of the fungi in Asia. Furthermore, the absence of the alternate host in Iraq, countries with the epidemiological zone 7 could be the source of spreading of new races of the fungi in Asia particularly Iraq, north eastern Syria, southeastern Turkey, Iran, northwestern Afghanistan and southwestern part of the former united soviet (Stubbs, 1988). It could explain why the spores were found during February to March while the disease does not appear on the susceptible cultivar SaberBeg until April.

Over the last decade a series of regional outbreak of yellow rust epidemics have been reported worldwide including Central and West Asia and East and North Africa (Ali et al., 2017). The nature of the disease attack has changed from an epidemic form to an endemic type in Iraq. High disease pressure has been observed annually since 2009 onward in most wheat growing area's especially in the years 2009, 2010, 2013, 2016 and 2019. During the last decade the most severe epidemics of yellow rust were detected during 2010 and 2019 where high levels of disease incidence and severities were recorded in most of the survived fields. In 2019, low to high levels of yellow rust infection incited by *P. striiformis* were found in 86.3% of the surveyed fields at various growth stages. Out of these 60.5% of the fields had low to high stem rust infection caused by *P. graminis* f. sp *tritici*, while leaf rust infection caused by *P. triticina* was found in 11.3% of the fields, and no infection was found in 1.3% of the fields. Wheat cultivars Buhoth 22, Barshlona, Cham 6, Al-Rasheed, IPA 99, Aras, Maxipak, SaberBeg and Abu Ghraib showed high levels of susceptibility to yellow rust disease in most farmer fields (Al-Maaroof, 2019).

5.5 Virulence Spectrum and Physiological Races of *Pst*

Composition of yellow rust population changes through time and this can be an important consideration for breeding programs. Monitoring for virulence changes in the pathogen population is very essential to aid in the development of resistant wheat cultivars. The pathogen has the ability to evolve rapidly into new races and to migrate long distances by air-borne dispersal (Chen et al., 2002). Limited work has been done on the variation in races of yellow rust prevalent in Iraq. Al-Baldawi (1993) mentioned that race 82E16 was the only existing race in wheat fields before the 1990 in Iraq, while the data obtained from the annual reports of the analysis of the physiological strains and their pathogenic ability in the third world countries indicated the prevalence of the 2E16 strains. The races 6E16, 70E16, 82E16 and 108E141 were reported in the Middle East region including Iraq (IPO, 1998). While Al-Maaroof et al. (2000) mentioned that the *Pst* races have the ability to infect the lines carrying resistance genes *Yr6, Yr7, YrA, Yr2, Yr9* and *Yr18*, in various combinations, which indicates the presence of more than one race in this region.

The severe yellow rust outbreaks in the middle zone of Iraq in 1998 resulted from acquisition of new virulence against *Yr9* gene, which was predominant in most of wheat cultivars as a single major gene (Al-Maaroof et al., 2004). Breakdown of *Yr9* resistance gene resulted in widespread epidemics in the Middle East and the Indian Subcontinent, which caused considerable crop damage in the 1990s (Singh, 2004). The origin of this series of pandemics is considered to be mutation to virulence for *Yr9* arising in the red sea region as early as 1987, areas which have been stimulated with the recent climatic condition changes (Louwers et al., 1992). Al-Maaroof et al. (2003) isolated and identified two physiological races of *Pst* for the first time in the wheat fields of middle Iraq. Race 6E16 was virulent on the known resistant genes *Yr6, Yr7, Yr8, Yr2, YrA*, and *Yr18* and the unknown resistant gene in Gereck 79 while race 230E150 showed its virulent on thirteen identified resistant genes (*Yr6, Yr7, YrSD, YrSu, Yr9 +, Yr7 +, Yr6 +, Yr8, Yr2 +, Yr2, YrA, Yr18* and *Yr9*) and one unidentified resistant gene in Gereck 79, and the strains 6E16 and 230E150 shared their ability to break the action of the diagnosed resistance genes *Yr6, Yr8* and *Yr9*. And *Yr18* and undiagnosed resistance gene present in the cultivar Gereck 79. The findings of the previous research study of the yellow rust pathogen physiological races suggested the prevalence of 6E16 race in the Middle East since 1973 (IPO, 1998). As reported in all of Syria, Lebanon, Turkey, Pakistan, Afghanistan and Saudi Arabia, from the continents of Asia, Egypt,Yemen, Tunisia, Morocco, Libya, Ethiopia, Algeria and Kenya from the continent of Africa during the period 1973-1998 (Hakim and El-Ahmed, 1998). The spread of this race in Iraq is therefore a natural consequence of its influx from neighboring countries and its spread throughout the region, although Al-Baldawi (1993) did not suggest its existence and confirmed that race 82E16 strain was the only race distributed in Iraq before the Nineteenth. While race 230E150 was recent in the region as it was registered in Kenya, Tanzania and Saudi Arabia during the time 1990-1992 (Lowers et al., 1992). It has occurred in Tanzania one year after its registration in Kenya. It is likely that the urediniospores of race 230E150 have passed to Iraq, transported by air currents winds coming from East Africa, represented by Tanzania, Ethiopia and Kenya, north to Southwest Asia, via Saudi Arabia, and then to the epidemiological region in which Iraq is located. The transition process may take place in several stages.

Regular field studies on virulence structure of *Pst* population were conducted annually using *Yr* biological trap nursery by the Iraqi rust surveillance team in coordination with ICARDA and BGRI. On the other hand, other NARC researchers also monitored the pathogen using the *Yr* differentials. Virulence's against the resistant genes *Yr6, Yr7, Yr9, Yr11, Yr18, YrA* and *YrSk* were detected in the natural population of *Pst* in Baghdad in 1998 and 1999 (Al-Hamdani et al., 2002; Al-Maaroof et al., 2003). While virulence against the known resistant genes, *Yr1, Yr6, Yr7, Yr9, Yr11, Yr15, YrSK, Yr18, Yr26*, and *Yr28* was common in the natural population of *Pst* in Sulaimani, north of Iraq in 2010, and no virulence was found against *Yr5, Yr8, Yr10, Yr12, Yr17, Yr24* (Al-Maaroof et al., 2009; AL-Bajalan, 2012). New virulence on the resistant genes *Yr25* and *Yr27* which caused a significant outbreak in many CWANA countries, was also identified in Iraq and caused serious epidemics in some wheat growing areas in 2010 (Al-Maaroof et al., 2011). The virulence spectrum of *Pst* has recently been highly evolved. The pathogen gained virulence against the known resistant genes *Yr2*, *Yr6*, *Yr7*, *Yr9*, *Yr18*, *YrA*, *Yr21*, *Yr25*, *Yr27*, *Yr28*, *Yr29*, *Yr31* at adult plant stage in the field in Sulaimani and against the known resistance genes *Yr1, Yr2*, *Yr4, Yr5*, *Yr7*, *Yr9*, *YrSD, YrSu*, *YrND, Yr32, YrA*, *Yr21*, *Yr27*, *Yr28*, *Yr31,* at adult plant stage in the field in Nineveh (Al-Maaroof et al., 2015). While virulence against the known resistant genes *Yr5*, *Yr6*, *Yr7*, *Yr9*, *Yr20*, *Yr21*, *Yr27*, *Yr28* and *Yr31 was common* in the natural populations of *Pst* in Babylon and against *Yr2*, *Yr5*, *Yr6*, *Yr7*, *Yr9*, *Yr18*, *YrA*, *Yr20*, *Yr25*, *Yr28*, *Yr29* and *Yr31* in Diyala (Al-Maaroof et al., 2020).

Race phenotyping analysis of *Pst* from the infected wheat leave samples collected from different infected wheat fields in Iraq at the Global Rust Reference Center (GRRC) of the

University of Aarhus, Denmark, for the period 2010 to 2016 revealed the predominance of two aggressive races in the natural *Pst* populations. The first race pathotype code is -, *Yr2*, -, -, -, *Yr6*, *Yr7*, *Yr8*, *Yr9*, -, -, -, -, *Yr27*, -, -, *AvS*, - while the second race pathotype code is -, *Yr2*, -, -, -, *Yr6*, *Yr7*, *Yr8*, *Yr9*, -, -, -, -, *Yr25*, *Yr27*, -, -, *AvS*, - (Hovmøller and Algaba, 2015; Al-Maaroof et al., 2020). Both races were aggressive based on the study of Millus et al. (2006), and tolerant to warm temperature. The findings of race phenotyping were confirmed by SSR genotyping of both races. The two races were clustered in the aggressive *PstS2* genetic lineage, the first race lineage variant called *PstS2* which was prevalent in South Asia, West Asia and East Africa and the second variants called *PstS2*, v27 which possess virulence against *Yr27*, which was widely distributed in East Africa, West Asia and South Asia (Walter et al., 2016; Ali et al., 2017). Both *Pst* races were predominant and responsible of the recent outbreak of the disease during 2009-16 which have associated with significant grain yield losses in most of the susceptible wheat cultivars and substantial economic loss in Iraq (Hovmøller et al., 2017). Race *PstS3* belonging to the *PstS3* genetic lineage was also detected in *Pst* populations in Iraq and some other west Asian and north African countries during the yellow rust epidemics in 2017 (Ali et al., 2017; Hovmøller et al., 2018). Indeed, Iraq is lying the Middle East, which has been shown to be diverse and is not very much distant from the pathogen centre of diversity and potential centre of origin (Ali et al., 2014; Ali et al., 2018). Future studies must be made to understand the population genetic structure of the pathogen using molecular markers and intensive surveillance efforts.

5.6 Genetic Resistance in Host Plants

Breeding for disease resistance is the most sustainable strategy to reduce yield losses, and is an environmentally safe approach to control the diseases (Singh and Trethowan, 2007). Great importance is given to improve disease resistance in wheat particularly to yellow rust disease using different breeding programs in Iraq (Ibrahim et al., 1993a, 1993b). Several resistant and high yielding wheat cultivars were released in the 1990s onward (NCFRRAV, 2014). Successive release of the resistant cultivars contributed in reducing grain yield losses caused by rust diseases (Al-Maaroof et al., 2001, 2005, 2012). With the introduction of the resistant varieties, new virulence and races of rust pathogens also appeared that have contributed to switching resistance of these varieties (Al-Maaroof et al., 2003, 2012, 2015). Therefore, breeding for resistance to rust diseases and developing new resistant cultivars having higher yield potential became the main target in all wheat breeding programs and considered as the most economical and effective way to eliminate the use of fungicides and reducing crop losses caused by the disease (Singh et al., 2004). Efforts have been made to establish genetic resistance to wheat rust diseases in Iraq by using nuclear techniques as an efficient method of inducing genetic variation in wheat by the use of different doses of gama rays and fast neutrons in the early 1980s, based on the successes achieved in the world in this area (Ibrahim and Al-Maaroof, 1986; Al-Ubaidi et al., 1989, 1993; Ibrahim et al., 1989a, 1989b, 1989c, 1989d, 1991, 2000; Al-Maaroof et al., 1989, 1993a, 1993b, 1995a; Al-Kubaisi et al., 1999). These efforts resulted in the inducing of a large number of mutant genotypes that showed different levels of resistance and susceptibility to yellow and brown rust diseases, either individually or in combination (Ibrahim et al., 1991, 1993b, 1993c, 1994; Al-Maaroof et al., 1993c, 2003b; Al-Ubaidi et al., 2001). The behavior of these genotypes was evaluated for rust diseases under artificial inoculation conditions at different environmental conditions compared to the local cultivars. Some resistant genotypes with high grain yield potential were developed into cultivars and approved by the National Committee for Registration and release of Agricultural Varieties at the Iraqi Ministry of Agriculture in order to spread their cultivation in the major wheat growing areas of Iraq (Adary, 1995; Ibrahim et al., 1998, 2000; Al-Jibouri et al., 2001; Al-Maaroof et al., 1998, 2001b, 2003d; Al-Ubaidi et al., 2000).

Some of the induced mutation cultivars, using fast neutrons explored different levels of resistance to yellow and brown rust diseases at the time of release, High resistance to yellow and brown rust diseases was identified in Tamuz 3 and resistance to brown rust disease was found in cv. Tamuz 2 (Ibrahim et al., 1994,1998), While gamma rays cultivars showed multiple resistance to yellow and brown rust diseases in Azmar 2 and Charmo (Al-Maaroof and Nori, 2018; Al-Maaroof et al., 2020), resistance to yellow and brown rust in Al-Kaed, Iratom and Intisar (Ibrahim et al., 1993a, 2001, 2002), and moderately resistance in Rabia and Al-Ize (Ibrahim et al., 2001a, 2001b).

The introduction of yellow rust resistance sources from the world wheat collections in a breeding program with the local wheat cultivars has contributed to an improvement in the genetic background for disease resistance in the local wheat cultivars and has led to develop the multi rust resistant cultivar Maaroof, and Charmo (Al-Maaroof et al., 2018, 2020) and the resistant cultivar Babel 113 (Al-Hamdani et al., 2005).

Other yellow rust resistant cultivars were obtained through the classical breeding programs and other novel technologies such as Farris1 and Reem (Al-Maaroof et al., 2009, 2012), Al-Nur and Hashemia (Al-Jibouri et al., 2000; Al-Janabi et al., 2001) and some other promising cultivars (Adary, 1995; Al-Ubaidi et al., 2002; Salman and Mahdi, 2005; Al-Maaroof et al., 2009b). Recently a wide array of commercial wheat cultivars in Iraq have succumbed to the new emerging *Pst* races. As a result, most of the new promising registered and released wheat cultivars have been subjected to a series of evaluation tests for the new races under high disease pressure at different farm locations. Various host parasite interaction has been established as shown in Table 5.1 (Al-Maaroof et al., 1995, 2000a, 2000b, 2003b, 2005, 2015, 2018; Al-Maaroof and Yahyawii, 2004).

While SSR markers have been widely applied in breeding programs for yellow rust resistance worldwide, there are still limited researches in Iraq on the identification of Yr resistance genes using molecular markers. Resistant genes *Yr7, Yr9, Yr18*, and *Yr26* have been widely found in most Iraqi wheat cultivars (Abed et al., 2011; Awad et al., 2017). Which may have an important role in the management of yellow rust diseases and is often associated with other rust genes. *Yr7* constantly associated with stem rust resistance gene *Sr9* and *Yr9* also associated with leaf rust resistance gene *Lr26* and *Sr31* and powdery mildew resistance gene Pm8 can provide durable resistance (Mclntosh et al., 1995). *Yr18* confer resistance at seedling and adult plant stages (Morgounov et al., 2012). While *Yr26* which is known in all Iraqi tested wheat cultivars and was effective resistance gene against many of *Pst* races, it can provide durable resistance with combination of other genes. Many *Yr* genes have been widely deployed in wheat genotypes by its association with leaf rust resistance or by its phenotyping effect on yellow rust response and considered to have durable yellow rust resistance (Singh, 1992).

Table 5.1 Origin, Pedigree and year of release of the registered Iraqi wheat cultivars and their resistant status to stripe rust disease on farm scale

No.	Cultivar Name	Synonym	Wheat Type*	Year of Release	Resistant Status	Pedigree	Breeding method	Origin
1	Saber Beg	Qandeharia	BW	n.a	HS	n. a	Selection	Local
2	Ajeeba	None	BW	1921	HS	Land race	Selection	India
3	Kenya Gular	Kenya, Greta	BW	1952	S	Land race	Selection	Australia
4	Florance Uror	Flouranca	BW	1958	S	Land race	Selection	France
5	Nouri70	None	BW	1958	S	CNO//SN64/KLRE/3/8156	-	-
6	Maxipak	Cite Ceross	BW	1966	S	Frantana/Kenya58/ Newthatch//Norin10 Bannt	Introduction	Mexico
7	Mexico 24	Maxikeya	BW	1967	S	Mexico 24	Introduction	Mexico
8	Inia 66	None	BW	1971	S	LR64/Son64/Lerma Rojo 64	Introduction	Mexico
9	Abu Ghraib	None	BW	1973	S	Ajeeba/Inia66/ Mexico24	Hybridization	Iraq
10	Tamuz 2	None	BW	1992	S	Sab. /Max. (Rad.)	Radiation(Nf)	Iraq
11	Tamuz 3	None	BW	1992	R	Sab. /Max.*AbuGh. (Rad.)	Radiation(Nf)	Iraq
12	Iratom 1	None	BW	1992	R	Sab. (Rad.)	Radiation (gamma)	Iraq
13	Ashur	AlTahadi, Taqa173	BW	1992	S	Sab./Max.	Hybridization	Iraq
14	Intsar	Iratom10	BW	1992	R	Sab./Lagis (Rad.)	Radiation (gamma)	Iraq

(continuous)

No.	Cultivar Name	Synonym	Wheat Type*	Year of Release	Resistant Status	Pedigree	Breeding method	Origin
15	AlMadain	Taqa 84 (AlNeda')	BW	1992	HS	Sab. /Max. (Rad.)	Radiation (gamma)	Iraq
16	Adnania	None	BW	1994	S	UP 114/ SaberBeg	Hybridization	Iraq
17	Sali	None	BW	1994	MR	Sab. /Lagis (Rad.)	Radiation (gamma)	Iraq
18	Rabia	1610	BW	1994	MS	Sab./HD (Rad.)	Radiation (gamma)	Iraq
19	Latifia	None	BW	1995	HS	Australian Line/Aras	Hybridization	Iraq
20	Iba'a 95	None	BW	1995	S	Veery"S"	Introduction	CIMMYT
21	Telfar 2	None	BW	1996	HS	Najah/Maxipak	Hybridization	Iraq
22	Telfar 3	None	BW	1996	S	Fath/Araz	Hybridization	Iraq
23	Al-Farris	Iba'a 99	BW	1997	MS	Ures/Bow"s"/3/Jup/ Biy "s" //Ures	Selection	CIMMYT
24	Al-Ize 66	None	BW	1997	S	Najah / Maxipak (Rad.)	Radiation (gamma)	Iraq
25	AlNour	None	BW	1997	HS	Car853/Coc/Vee/3/Bow	Introduction	ICARDA
26	Al-Hashmia	None	BW	1997	R	BR12*3/3Br14/LD*6	Introduction	ICARDA
27	Al-Iraq	AlKaed278	BW	1999	S	Max.(Rad.)	Radiation (gamma)	Iraq
28	Babel 113	None	BW	2000	MR	Max. /R23	Hybridization	Iraq
29	Cham 4	Sham4	BW	2000	MS	W-3918-A/JUPATECO-73-	Introduction	ICARDA
30	Cham 6	Sham6	BW	2000	S	PLC"S" - Ruff"S"/Gta"S" -RTTE CM-12904-1M-3M-1Y-1Y-OSK-OAP	Introduction	ICARDA
31	Ur	None	BW	2000	S	ID CM 78108-IM-02Y-02M-16Y-2B-OY)	Introduction	ICARDA
32	Al-Melad	None	BW	2000	S	—	Selection	Iraq
33	Sawa	AlNakhwa	BW	2000	—	SaberBeg/Maxipak (Rad.)	F2 Radiation	Iraq
34	Al-Rashid	Al-Mansur Bellah	BW	2001	S	Max. (Rad.)	Radiation (gamma)	Iraq
35	Al-Hadba'	None	BW	2001	S	SABERBEG/ Inia 66 (Rad.)	Radiation (gamma)	Iraq
36	Al-Fateh	None	BW	2001	S	CI8224/CI6833/Conz/7IIC/ Top-Swm6328/9AP-OAP// Max.	Hybridization	Iraq
37	Rizgari	None	BW	2010	S	Cross Bread 11-	Introduction	
38	Buhoth 22	None	BW	2011	S	CMSS96Yo3204F-PJN/ Bow//OPATA	Introduction	CIMMYT
39	Furat	None	BW	2011	S	5H-840964HSI	Introduction	—
40	Farris 1	Farris	BW	2012	MS	STAR/TR771773/SLM	Selection	CIMMYT
41	Dijla	None	BW	2012	HS	8409644HS2-6H	Introduction	USA

(continuous)

No.	Cultivar Name	Synonym	Wheat Type*	Year of Release	Resistant Status	Pedigree	Breeding method	Origin
42	Adana 99	None	BW	2012	S	PFAU/SERI-M-82/ BOBWHITE	Introduction	Turkey
43	Baghdad 1	None	BW	2012	S	MX105-6MVLT40/BNSN	Introduction	CIMMYT
44	Buhoth 158	None	BW	2012		118//S2/57-S2-CR7-S2	Hybridization	Local
45	Al-Baraka	hh	BW	2013	S	IARI*STD	Introduction	Iraq
46	Buhoth 4	None	BW	2013	MS	S201*	Introduction	—
47	Buhoth 10	None	BW	2013	R	Iba' a 95/Iba' a 99	Hybridization	Local
48	Uruk	None	BW	2013	MR	Inia 66 (Rad.)	Radiation (gamma)	Iraq
49	Cham 7	Sham 7	BW	2013	—	—	Introduction	ICARDA
50	Bengal	None	BW	2013	HS	BISU/3/YAV79/ALO1/ ALTARS4/ CD93683. 7Y. 040M-03OY-LPAP. B	Introduction	Spain
51	Al-Maaroof	None	BW	2014	R	KTN M12/ DAMANxADL//TAMUZ 2	Hybridization	Iraq
52	Alaa	None	BW	2014	MR	ESDA/VEE*10	Selection	Iraq
53	Hsad	None	BW	2014	MSS	SNB// CMH79A955/3*CNO79/3/ ATTILA	Selection	Iraq
54	Bura	None	BW	2014	S	H31/Trapf21/Enesco	Introduction	Italy
55	Baghdad 3	None	BW	2014	MR	—	Selection	Iraq
56	Al-Rashedia	Beaj3	BW	2014	S	TRACHAS//CMH76252/ PVNS/78	Introduction	—
57	Al-Fiahd	None	BW	2014	—	—	Selection	
58	Siktero	None	BW	2014	—	Adam/2282	Introduction	France
59	Dijlat AlKhair	Hesio	BW	2015	—	Orpic*/isengrain)* isengrain	Introduction	France
60	Nogal	None	BW	2015	—	—	Introduction	France
61	Wafia	None	BW	2015	S	—	Introduction	France
62	Al-Hana'	None	BW	2015	—	Ajeeba (Rad.)	Radiation	Iraq
63	Radhia	None	BW	2015	—	—	Introduction	France
64	Aras	None	BW	2016	S	Sonora 64/ Lerma (Rojo 64) // Sentaclena	Selection	Iraq
65	Slemani 2	Kauz*2	BW	2016	S	Kauz*2	Introduction	CIMMYT
66	Ceyhan99	Jyhan99	BW	2016	S	BJY'S'/COC	Introduction	Turkey
67	Dania	A*A	BW	2016	—	Dahbi6/3/Ardi1/Topo/419// Eriz09/4/Eri2011*2/Milan	Selection	-
68	Hawler 2	None	BW	2018	S	Florkwa-2/6/Sakers/5// YMH/TOB/4/BOWS- LC96-0180-030APS-2AP- DAPS-OPA	Introduction	CIMMYT

(continuous)

No.	Cultivar Name	Synonym	Wheat Type*	Year of Release	Resistant Status	Pedigree	Breeding method	Origin
69	Hawler 4	None	BW	2018	MSS	Baj#1/3/KirijAI/A*2/ PastorCMSSOYOO288S-OS8-099Y-99M-099Y3M-OWGY	Introduction	CIMMYT
70	Ding	None	BW	2018	—	Kauz*Pastor	Introduction	Turkey
71	Kalar1	None	BW	2018	S	ID800994.W/VEE/5/ CA8055/4/ROMTAST/ BON/3DIBO//SU	Introduction	ICARDA
72	Kalar2	None	BW	2018	S	CH75479/SARDARI-HD74	Introduction	ICARDA
73	Aldiyar	None	BW	2019		K58w418/SPN/3/CHEN/ AE.SQ//2*OPATA/4/ FREr2/2* SOKOLL/PAS ro TOR/HXL7573/2*BAU	Introduction	CIMMYT
74	KM5180	None	BW	2019		—	Selection	—
75	AlMahmoodia	None	BW	2019		SOKILL/3/PASTOR// MILAn/3/BAV92	Introduction	CIMMYT
76	Azmar 2	None	BW	2020	R	CMH 83/ ELVIRA// CMH 79/ AGA/ INIA (Rad.)	Radiation (gamma)	Iraq
77	Charmo	None	BW	2020	MR	CMH79A/AGA/ CNO67//INIA66/NAC/ BABAX(Rad.)	Radiation (gamma)	Iraq
78	Sorah Gool	None	DW	n.a	—	Land race	Selection	local
79	Semakiya	None	DW	n.a	S	Land race	Selection	local
80	Falastinia	Karonia	DW	n.a	—	Land race	Selection	local
81	Senator Kapelli	Italia	DW	1951	S	Land race	Selection	Itali
82	Jori	None	DW	1970	—	BYE*2/4*TC60// TAC125E/3*TC60	Selection	Mexico
83	Sin el Jamal	None	DW	n.a	S	Land race	Selection	local
84	Izra	None	DW	n.a	—	—	—	—
85	Cocoret	None	DW	1970	—	RAE/4*TC60//STW63/3/ AA “S”	Selection	Mexico
86	Jerardo	None	DW	n.a	S	—	—	—
87	Iba’a 131	None	DW	1995	MS	POL “S”/Caborca 79	Selection	
88	Wahat el Iraq	None	DW	1995	R	Pics Ruffs/Rtte/Gta	Introduction	
89	Um Rabia	None	DW	1997	MR	Eross Joric 69/Hau	Introduction	
90	Dor 29	Babyl 29	DW	2000	MS	Sin el Jamal /Izra131	Hybridization	Iraq
91	Taqani 1	None	DW	2000	—	—	Introduction	
92	Al-Ibrahemia	None	DW	2001	—	Sab. /Sin el Jamal	Hybridization	Iraq
93	Al-Naama	Dor85	DW	2002	S	Izra 131/ Cocoret 71	Hybridization	Iraq
94	Sardar	None	DW	2010	MR	Cyprus	Introduction	Cyprus

(continuous)

No.	Cultivar Name	Synonym	Wheat Type*	Year of Release	Resistant Status	Pedigree	Breeding method	Origin
95	Buhoth7	None	DW	2010	—	EDM/KiaCYD-84-330-29D-OP-614-OP	Introduction	Cyprus
96	Simeto	Cimmito	DW	2010	MR	Capeiti 8/ Valnova	Introduction	Italy
97	Crisso	None	DW	2010	S	CP B14(Yt 54-N10-B)/ Cp63)/TC60	Introduction	Italy
98	Ofanto	Ovanto	DW	2011	S	Adamello/ Appulo	Introduction	Italy
99	Grecale	None	DW	2012	—	—	Introduction	Italy
100	Svevo	None	DW	2012	—	CIMMY Selections/Zinet	Introduction	Italy
101	Acsad65	Acsad	DW	2013	S	SNB// CMH79A955/3*CNO79/3/ ATTILA	Introduction	CIMMYT
102	Douma	None	DW	2013	MS	—	Selection	
103	Latifia2	None	DW	2013	S	—	Selection	CIMMYT
104	Iride	None	DW	2014	—	Altar 84/Ionio	Introduction	Italy
105	Mikki 3	None	DW	2016	R	—	Introduction	ICARDA
106	Hawler3	None	DW	2018	S	—	Introduction	

5.7 Other Disease Control Measures

Chemical control may be considered as the first line of defense against the disease worldwide. In Iraq chemicals are rarely applied to control the disease on large scale particularly in the low input farms. Usually, the chemicals are applied to protect susceptible cultivars from the early infection especially on the flag leaves. A series of chemicals were recently applied to control yellow rust disease in the high input farms. Al-Maaroof (2014) reported that application of the recommended dose of FS720 at the initial stages of the yellow rust onset resulted in a full control of the yellow and brown rust disease at a rate of 100 per cent, which reflected positively the increase of the grain yield by 40% in SaberBeg and 23% in Tamuz 2 compared to the control treatment (Al-Maaroof, 2014). While Al-Maaroof and Nori (2014) indicated that two applications of Bayfidan 250EC (Traidemnol) at the rate of 0.5L/ha completely controlled yellow rust development on different susceptible wheat cultivars.

Induced disease resistance which activates the natural plant defense, have been applied as a tool for yellow rust resistance. It is environmentally safe and confers long-lasting resistance against the disease. Application of IAA, BTH, BABA and SA significantly reduced disease severity, coefficient of infection and rate of infection in Tamuz-2. and AL-8/70. However, different cultivars reacted differently toward the chemical inducers. Changes in peroxidase, phenylalanine ammonia-lyase activities and the total phenolic compound content were associated with resistance against *Pst* in both cultivars by the inducers (Fayadh et al., 2013), Foliar application of proteck MgS^{+} at a concentration of 3ml l^{-1} significantly reduced disease incidence of yellow rust disease on the susceptible wheat cultivars, disease reduction was associated with the increases in peroxidase and phenylalanine ammonia-lyase enzyme activity which increased the accumulation of phenolic compounds (Shams-Allah and Hussein, 2020).

Cultural practices provide another method for at least partial control of wheat rust epidemics. No single practice is effective under all conditions, but using a series of cultural practices greatly enhances the existing resistances. Al-Maaroof (1997) recommended to sow wheat cultivar early in November to avoid the early infection and escape from the disease at early flowering stage and spikes formations.

Variety mixture technique is another economical and effective tool in reduction of rust disease in Iraqi wheat field. Al-Maaroof (1997) reported the possibility of using variety mixtures in the north and middle of Iraq to reduce impact of wheat yellow rust. It was also found that variety mixture can reduce the infection rate and AUDPC of yellow rust disease by 48%~55% and by 38%~44% respectively over the means of the pure stand. Moreover, the mixtures had performed better than their pure stand, with mean yield increased by 4.7%~11.7% in the presence of yellow rust in rainfed area in the northern Iraq. Variety mixtures has no significant effect on the quality and bread making ability of the mixed flour (Al-Maaroof, 1997). Results of using three-way mixtures revealed that the mean disease severity in the mixture was reduced compared to the pure stands components between 25%~79%. Some mixture yielded 18% more than mean of the pure stands in the presence of the disease. Meanwhile, varietal mixture resulted in an increase in yield components, with 0.4%~8.2% increase for thousand kernels weight and 0.3%~8.5% for grain number per spike. Furthermore, some mixtures yielded as high as or higher than the best cultivar in pure stands in the presence of yellow rust (Al-Maaroof et al., 2000). Three-way mixtures were also effective in reducing brown rust severity and rising yield components per cultivars and a subset of the mixtures and pure stands. The mean disease severity in the mixture compared to their components grown in pure stands were reduced between 11%~63%. The reduction was reflected in increasing of the specific weight of the grain in the mixtures (Al-Maaroof et al., 2004, 2006).

5.8 Prospective

Continuous monitoring and forecasting of the prevailing *Pst* races from the main Iraqi wheat growing areas is very much necessary, which can save time and speed up the breeding programs for disease resistance. The current virulence spectrum of *Pst* should assist in pyramiding resistance genes and their deployment in the national breeding program. Postulation of the *Yr* genes in all the Iraq wheat cultivars is also very much important. As a center of wheat origin, exploring the sources of yellow rust resistance genes in wheat landraces and the wild wheat relatives which may be lost during wheat domestication and intensive breeding program and re-introduce them in the commercial wheat cultivars will be another promising option to strengthen the breeding for resistance to the diseases. Furthermore, focusing on the molecular genetic diversity of the pathogen is very much important. The information should be helpful in understanding the pathogen population and to devise control measures to minimize the amount of yield losses caused by the disease in Iraq.

References

Abed HA, Ogbonnaya F, Nazari K, et al., 2011. Molecular Screening of Iraqi Wheat Cultivars for Rust Diseases. ICARDA Training Report Application of Biotechnology for Crop Improvement.

Adary AH, 1995. Development of the Bread Wheat Cultivar "Adnanya" for the Limited and Moderate Rain-fed Area of Northern Iraq. *IPA J. Agric.*, 5(1):1-10. Al-Adami AR., 1953. A Preliminary List of Plant Disease in Iraq. Bulletin Ministry of Agriculture, Iraq, 17:1-14.

Al-Bajalan AM, 2012. Studies on Some Economic Importance and Epidemic Aspects on Wheat Yellow Rust in Sulaimani, Iraq. MSc. Thesis, College of Science, University of Sulaimani, Iraq.

Al-Baldawi AA, 1993. Occurrence and Importance of Wheat and Barley Diseases in Iraq. In: Proceedings, Technology Transfer in the Production of Cereals and Legumes Workshop, September 20-22, 1993, Mosul, Iraq :105-113.

Al-Baldawi AA, 1981. Susceptibility of Some Wheat Variety to Leaf Rust Disease. *Yearly Book of Plant Protection Researches*, 2(2):37-47.

Al-Chaabi S, Abu-Fadel T, 2012. Epidemic incidence of yellow rust disease on bread wheat in Syria During 2010 Season, Performance of Released and Promising Varieties, and Preliminary Detection of Remained Effective Resistant Genes to Pathogen Virulence's. *Arab Journal of Plant Protection* , 30:180-191.

Al-Hamdany MA, Abas HA , Khadem AH, 2002. Shifting in the Virulence Pattern of Wheat Leaf Rust Causal Agent in Baghdad Region. *Iraqi Journal of Agriculture*, 7:9-17.

Al-Hamdany MA, Kadhem AH, Abass HA, et al., 2005. Development of Bread Wheat Cultivar "BABIL 113" in Iraq. *Iraqi Journal of Agriculture*, 7: 9-17.

Al-Hamdany MA, Wellings CR, Kadhem AH, et al., 2002. Virulence Pattern of Yellow Rust Causal Agent Population *Puccinia striiformis* Westend. on Near Isogenic Lines in Baghdad Area. *Arab J. Plant Protection*, 20(1):24-28.

Ali S, Gladieux P, Leconte M, et al., 2014. Origin, Migration Routes and Worldwide Population Genetic Structure of the Wheat Yellow Rust Pathogen *Puccinia striiformis* f. sp. *tritici*. *PLoS Pathogen*, 10 (1): e1003903. doi:10.1371/journal.ppat.1003903.

Ali S, Rodriguez-Algaba J, Thach T, et al., 2017. Yellow Rust Epidemics Worldwide Were Caused by Pathogen Races from Divergent Genetic Lineages. *Front Plant Sci*., 20(8):1057. DOI: 10.3389/fpls.2017.01057. PMID: 28676811; PMCID: PMC5477562.

Ali S, Sharma S, Leconte M, et al., 2018. Low Pathotype Diversity in a Recombinant *Puccinia striiformis* population through Convergent Selection at the Eastern Part of Himalayan Centre of Diversity (Nepal). *Plant Pathology*, 67:810-820.

Ali S, Rodriguez-Algaba J, Thach T, et al., 2017. Yellow Rust Epidemics Worldwide were Caused by Pathogen Races from Divergent Genetic Lineages. *Frontiers in Plant Science*, 8:1058.

Al-Janabi KK, Al-Maaroof EM, Yousif DP, 2001. Introduction Genotypes to Induce New Cultivars of Bread and Durum Wheat, Triticale and Barley Suitable for Iraqi Environments. *Al-Buhooth Al- Tachaniya J.* , 14 (76):124-136.

Al-Jibouri AJ, Jaddou KA, Al-Janabi KK, et al., 2000. Production of Two Bread Wheat Cultivars for Irrigated Lands of Iraq. *Arab Agric. Res. J.* , 4(2):178-197.

Al-Kubaisi NM, Ibrahim IF, Al-Maaroof EM, et al. , 1999.Induced New Wheat Mutants by Gamma Rays. *Diala Journal* , 7: 47-53.

Al-Maaroof EM, 1997. The Role of Variety Mixture in Disease Control of Wheat Rusts Caused by *Puccinia recondita* and *P. striiforims* in Iraq. Ph. D Dissertation, College of Agriculture, University of Baghdad.

Al-Maaroof EM, 2014. Efficiency of FS-720 in Disease Control of Wheat Rusts Diseases. A report submitted to the national committee for registration and approval of Pesticides. MOA, Baghdad, Iraq.

Al-Maaroof EM, 2017. Identification of Physiological Races of *Puccinia graminis* f. sp. *tritici* in Iraq. *Journal of Wheat Research*, 9:47-53.

Al-Maaroof EM, 2021. Occurrence and Virulence's of Wheat Yellow Rust Disease in Iraq. In: ProcSeedings, The International Conference of the Sustainable Agricultural and its Role in Human and Economic Development, Feb.17-18, 2021, Basrah, Iraq.

Al-Maaroof EM, Abass KK, Abdullah SH, et al., 2009a. "Reem" A New Wheat Cultivar Resistant to Brown and Yellow Rust Diseases and with High Yield Potential. *Iraq Journal of Agriculture*, 14:165-175.

Al-Maaroof EM, Abass KK, Fiahd FA, et al., 2012. Developing of New Wheat Cultivar "Farris1" with High Yield Potential and Resistance to Yellow and Brown Rust Diseases. *Arab Journal of Plant Protection*, 30:213-222.

Al-Maaroof EM, Ahmed MY, Hussein WU, 2003a. Virulence of Wheat Yellow Rust Disease in Iraq. *Iraqi Journal of Biology*, 13:1-10.

Al-Maaroof EM, Al-Ani RA, Latif MM, et al., 2004. Brown Rust Disease Control in Wheat by Variety Mixtures.

Iraqi Journal of Agriculture Science, 5:97-101.

Al-Maaroof EM, Al-Baldawi AA, Aboud AR, et al., 2000b. Response of Registered and Released Wheat Cultivars Against Leaf Rust Disease Caused by *Puccinia recondita* in Iraq. *Iraqi Journal of Agriculture*, 5:110-120.

Al-Maaroof EM, Fayadh AH, Hovmoller M, 2020. Pathogenic Divergence in *Puccinia striiformis* f. sp. *tritici* Populations, The Causal Agent of Yellow Rust Disease of Wheat in Iraq. Proc. of the BGRI Technical Workshop.

Al-Maaroof EM, Fiahd FA, Abdullah SH, et al., 2009b. Improving Brown Rust Resistance in Wheat Cultivar Tamuz 2. Proc. 5th Int. conf. of Plant Pathology Nov.10-13, New Delhi, India.

Al-Maaroof EM, Hadwan HA, Mohamed LO, et al., 2012. Serious Outbreak of Wheat Yellow Rust Disease in Iraq. *Journal of University of Duhok*, 15:358-369.

Al-Maaroof EM, Hakim MS, Yahyaoui A, 2003. Isolation and Identification of Tow Physiological Races of Wheat Yellow Rust Pathogen *Puccinia striiformis* West f. sp. *tritici* in Iraq. *Iraqi Journal of Agriculture Science*, 34:157-164.

Al-Maaroof EM, Hovmøller M, Ali RM, et al., 2015. Detection of *Yr27* Virulence in *Puccinia striiformis* f. sp. *tritici* Population on Wheat in Iraq. *Journal of Wheat Research*, 7:39-47.

Al-Maaroof EM, Ibrahim IF, Aboud AR, 2001a.Effect of Leaf Rust Disease *Puccinia recondita* Rob. ex Desm. f. sp. *tritici* on Different Wheat Cultivars and Genotypes in Iraq. *Dirasat Journal of Agriculture Research*, 28:111-120.

Al-Maaroof EM, Ibrahim IF, Abass HA, et al., 2000. Use of Variety Mixtures in Control of Wheat Yellow Rust in Iraq. *Arab Agricultural Research Journal*, 4:198-220.

Al-Maaroof EM, Ibrahim IF, Aboud AR, 2000a. Effect of Leaf Rust Disease on Grain Yield and Quality of Hard Wheat. In: The Proceeding of the 7th Congress of Foundation of Technical Institute, Baghdad, Iraq.

Al-Maaroof EM, Ibrahim IF, Al-Ubaidi MO, et al., 1993. Induced Resistant Mutants to Wheat Leaf Rust by Gamma Rays. In: Proceedings, 4th Scientific Conference for the High Council of Scientific Societies. Oct. 26-28. Babylon, Iraq.

Al-Maaroof EM, Ibrahim IF, Al-Janabi AA, 1995. Host Reaction of Some Wheat Cultivars with *Puccinia Recondite* Rob. ex Desm. f. sp. *tritici* in Iraq. *Arab Journal of Plant Protection*, 13:86-89.

Al-Maaroof EM, Ibrahim IF, Kraibit AA, 2001b. Response of Some Bread Wheat Genotypes to Yellow Rust Disease Caused by *Puccinia striiformis* in Iraq. *Arab Journal of Plant Protection*, 19:12-18.

Al-Maaroof EM, Latif MM, Nefal AA, 2006. Variety Mixture as a Tool for Brown Rust Disease Control in Wheat. Proceedings of the 12th congress of the Mediterranean Phytopathological Union, 11-15, June, Rhodes Island, Athens, Greece.

Al-Maaroof EM, Latif MM, Said HA, et al., 2002a. Detecting the Effective Resistant Genes to Leaf Rust Disease *Puccinia recondita* Rob. ex Desm. f. sp. *tritici* on Wheat in Iraq. *Arab Journal of Plant Protection*, 20:157-

164.

Al-Maaroof EM, Nazari K, Hodson D, et al., 2011. Monitoring and Distribution of Yellow Rust Disease on Wheat in Iraq. Proceedings of the International Wheat Yellow Rust Symposium, April 18-20, 2011, ICARDA, Aleppo, Syria.

Al-Maaroof EM, Nori AM, 2018. Yellow Rust Development on Different Wheat Genotypes. *Journal of Zankoy Sulaimani Part-A*, Special issue:177-188.

Al-Maaroof EM, Saleh RM, Mahmood HA, et al., 2020. Developing the New Multi rust Resistant Bread Wheat Cultivar "MAAROOF" for the Irrigated and Rain-fed Zones of Iraq. *Applied Ecology and Environmental research*, 50:814-826.

Al-Maaroof EM, Singh R, Hussein AK, et al., 2003b. Host Reaction of Some Wheat Genotypes with *Puccinia striiformis*, the Causal Agent of Yellow Rust Disease in Iraq. *Iraqi Journal of Agriculture*, 8:70-78.

Al-Maaroof EM, Singh RP, Huerta J, et al., 2005. Resistance of Some Iraqi Bread Wheat Cultivars to *Puccinia triticina*. *Phytopatholgia Mediterrania*, 44:247-255.

Al-Maaroof EM, Yahyaoui A, 2004. Response of Some Wheat Genotypes to Yellow and Leaf Rust Diseases. *Iraqi Journal of Agriculture Science*, 5:15-20.

Al-Mashhadani AH, 2014. Genetic Variation in *Puccinia striiformis* Westend f. sp. *tritici* Population, the Causal Agent of Wheat Yellow (yellow) Rust and Some Control Measure in Iraq. Ph.D. Dissertation, College of Agricultural Engineering Sciences, University of Sulaimani.

Al-Ubaidi MO, Ibrahim IF, Al-Maaroof EM, 2001. Induced Mutants in Durum Wheat Resistant to Brown Leaf Rust Disease by Gamma Rays. *Arab Agric. Res. J*, 5(1):78-89.

Al-Ubaidi MO, Ibrahim IF, Al-Maaroof EM, 2002. Development of New Durum Wheat Cultivars Resistant to Brown Leaf Rust by Gamma Rays. *Ibn Al-Haitham J. for Pure & Appl. Sci.*, 15:1-10.

Awad SR, Fayadh AH, Al-Mawlood MA, 2017. Identification of Some Stem and Yellow Rust Resistance Genes in Some Iraqi Varieties by Using PCR Markers. *Iraqi J. Agric. Res.*, 22(1):192-199.

Baker F, 1973. Methyl Mercury Poisoning in Iraq. *Science*, 181:241-280.

Beddow JM, Pardey PG, Chai Y, et al., 2015. Research Investment Implications of Shift in the Global Geography of Wheat Yellow Rust. *Nat. Plants*, 1:15-32. DOI: 10.1038/nplants.2015.132.

Bernardo R, 2002. Breeding for Quantitative Traits in Plants. Minnesota: Stemma Press.

Chader AA, Al-Janabi, Kraibit AA, 1998. Induced New Wheat Cultivar Resistant to Brown Leaf Rust Disease by Fast Neutrons in Iraq. In: Proceeding of the Fourth Arab Conference on the Peaceful Uses of Atomic Energy. Nov. 14-18, Tunis.

Chakraborty S, Luck J, Hollaway G, et al., 2010. Rust Proofing Wheat for a Changing Climate. In: Proceedings of Borlaug Global Rust initiatives 2010 Technical Workshop, Russia.

Chen XM, 2005. Epidemiology and Control of Yellow Rust *Puccinia striiformis* f. sp. *tritici*, on Wheat. *Can. J. Plant Pathol.*, 27:314-337.

El-Naimi M, Mamluk OF, 1995. Occurrence and Virulence of Wheat Rust in Syria. *Arab J. Plant Protection*, 13(2):76-82.

FAO, 2018. Global Information and Early Warning System on Food and Agriculture.

FAOSTAT, 2018. World Food and Agriculture, Statistical Pocket Book.

Fayadh A, Al-Maaroof EM, Fattah F, 2013. Induced Resistance to Wheat Yellow Rust by Chemical Inducers. *Journal of Biology, Agriculture and Healthcare*, 3(20): 56-63.

Hakim MS, El-Ahmed A, 1998. The Physiological Races of Yellow Rust *Puccinia striiformis* f. sp. *tritici* in Syria During the Period 1994-1996. *Arab J. Plant Protection*, 16:7-11.

Hawkesford MJ, Araus JL, Park R, et al., 2013. Prospects of Doubling Global Wheat Yields. *Food and Energy Security*, 2:34-48.

Hovmøller MS, Algaba JR, 2015. Report for *Puccinia striiformis* Race Analyses 2014, Global Rust Reference Center (GRRC), Aarhus University, Flakkebjerg, DK- 4200 Slagelse, Denmark.

Hovmøller MS, Algaba JR, Thatch T, et al., 2017. Report for *Puccinia striiformis* Race Analyses and Molecular Genotyping 2016, Global Rust Reference Center (GRRC), Aarhus University, Denmark.

Hovmøller MS, Algaba JR, Thatch T, et al., 2019. Report for *Puccinia striiformis* Race Analyses and Molecular Genotyping 2018, Global Rust Reference Center (GRRC), Aarhus University, Denmark.

Ibrahim IF, Al-Maaroof EM, 1986. Resistance of M2 Selections of Wheat Induced by Fast Neutrons to Leaf Rust Disease. The 4th. Sci. Conf. of Sci. Res. Council, 1:1520-1524.

Ibrahim IF, Al-Maaroof EM, Al-Aubaidi MO, et al., 2001a. Induction of New Bread Wheat Cultivar "Rabia" Resistant to Yellow and Brown Rust Diseases to Rain-fed Areas. *Iraqi J. for Agric. Sci.*, 2(2):82-87.

Ibrahim IF, Al-Maaroof EM, Al-Aubaidi MO, et al., 2002. Induction of New Bread Wheat Cultivar "Al-Kaed" Resistant to Yellow and Brown Rust Diseases by Gamma Rays. *IPA. J. of Agr. Res.*, 12(1):11-25.

Ibrahim IF, Al-Maaroof EM, Ghader MO, et al., 1993. Induction of New Wheat Varieties Resistant to Leaf Rust and with Good Agronomic Traits by Nuclear Techniques. In: The Proceedings of the Workshop on Induced Mutation in Plant Improvement. AAEA. Nov.1-3, 1993, Baghdad, Iraq.

ICARDA, 2013. Improved Livelihoods of Small Farmers in Iraq through Integrated Pest Management and Organic Fertilization-Final report, IRAQ-ICARDA-IFAD PROJECT (IFAD GRANT NO. 1001-1Q).

ICARDA-HSAD, 2014. Harmonized Support for Agricultural Development in Iraq-Final Report, USAID Cooperative Agreement No. AID- 267-IQ-12-0001.

IPO, 1998. The Research Institute for Plant Protection, 1972-1998. Reports on the Monitoring Pathogenicity Patterns of Yellow Rust on Wheat in the Third World.

Jaradat AA, 2002. Agriculture in Iraq: Resources, Potentials, Constraints, and Research Needs and Priorities. A report submitted to department of State–Middle East working group on agriculture October 5-6, 2002, Washington D.C., USA.

Jin Y, Szabo LJ, Carson M, 2010. Century-Old Mystery of *Puccinia striiformis* Life History Solved with the

Identification of *Berberis* as an Alternate Host. *Phytopathology*, 100:432-435.

Karakas O, Hasancebi S, Ertugrul F, et al., 2009. Est-Based Multiplex Gene Expression in Yellow Rust Infecte Wheat Using Genome Lab Gexp Genetic Analysis System. In: Proceedings of 12th international cereal rust and powdery mildews in Antalya, Turkey, 13-16 October.

Kharouf S, 2009. Study on the Regional Genetic Variation of Wheat Yellow Rust *Puccinia striiformis* f. sp. *tritici*, Using DNA Molecular Markers. Ph. D Dissertation, University of Damascus, Syria.

Kumarse N, 2011. Status of Wheat Yellow Rust in CWANA: Analysis of Current Outbreaks. In: Proceedings of International Wheat Yellow Rust Symposium, Aleppo, Syria.

Lowers JM, Van Silfhout CH, Stubbs RW, 1992. Race Analysis in Wheat in Developing Countries, Report 1990-1992.

McIntosh RA, Wellings CR, Park RF, 1995. Wheat Rusts- An Atlas of Resistance Genes. CSIRO Publications, Australia.

Milus EA, Seyran E, McNew R, 2006. Aggressiveness of *Puccinia striiformis* f. sp. *tritici* Isolates in the South-Central United States. *Plant Disease*, 90:847-852.

MOA Representative, 2020. Iraq Reached Self-sufficiency in Wheat Production. Shafaq News, 12/8/2020, https://shafaq.com.

Morgounov A, Tufan HA, Sharma R, et al., 2012. Global Incidence of Wheat Rusts and Powdery Mildew during 1969-2010 and durability of Resistance of Winter Wheat Variety Bezostaya I. *Eur. J. Plant Pathol*., 132:323-340.

Nazari K, 2011. Status of Wheat Yellow in CWANA Analysis of Current Outbreaks. International Wheat Yellow Rust Symposium, ICARDA, Aleppo, Syria, 18-20, April.

NCFRRAV, 2014. Agricultural Cultivars and Hybrids Database. National Committee for Registration and Release of Agricultural Varieties/Iraqi Ministry of Agriculture.

Newton AC, Pretorius S, George P, 2010. Implications of Climatic Change for Diseases, Crop Yields and Food Security. In: Proceedings of Borlaug Global Rust initiatives 2010 Technical Workshop, Russia.

Rapilly F, 1979. Yellow Rust Epidemiology. *Ann. Rev. Phytopath*., 17:59-73.

Roelfs AP, Bushnell WR, 1985. The Cereal Rusts Vol. 2. Disease Distribution Epidemiology and Control. Orlando: Academic Press Inc.

Roelfs AP, Singh RP, Saari EE, 1992. Rust Disease of Wheat, Concepts Methods of Disease Management. Mexico, D.F, CIMMYT.

Salman RM, Mahdi AS, 2005. Selecting New Promising Lines of Bread Wheat. *The Iraqi Journal of Agricultural Sciences*, 36(5):67-74.

ShamsAllah SA, Hussien HS, 2020. Efficacy of Proteck MgS+ in Reducing Yellow Rust Disease on Wheat Caused by *Puccinia striiformis* f. sp. *tritici*. *Indian Journal of Ecology*, 47:143-145.

Singh RP, 2004. The Cost to Agriculture of Recent Changes in Cereal Rusts. In: Proceedings of the 11th

International Cereal Rusts and Powdery Mildews Conference, John Innes Center, Norwich UK, August.

Singh RP, Duvillier E, Huerta-Espino J, 2004. Virulence to Yellow Rust Resistance Gene *Yr27*. In: A new threat to stable wheat production in Asia. (Abs.). Second Regional Yellow Rust Conference for CWANA, Islamabad, Pakistan, 22-26 March.

Singh RP, Trethowan R, 2007. Breeding Spring Bread Wheat for Irrigated and Rain-fed Production Systems of the Developing World. In: Kang M, Priyadarshan PM (eds.), Breeding major food staples. Iowa: Blackwell Publishing.

Singh RR, 1992. Genetic Association of Leaf Rust Resistance Gene *Lr34* with Adult Plant Resistance to Yellow Rust in Bread Wheat. *Phytopathology*, 82:835-838.

Stubbs RW, 1988. Pathogenicity Analysis of Yellow Rust of Wheat and its Significance in a Global Context. In: Breeding strategies for resistance to the rusts of wheat. Simmonds NW and Rajaram S (Editors). CIMMYT, MEXICO, D. F.

Torabi M, Mardoukhi V, Nazari K, et al., 1995. Effectiveness of Wheat Yellow Rust Resistance Genes in Different Part of Iran. *Cereal rust and powdery mildews bulletin*, 23:9-12.

Walter S, Ali S, Kemen E, et al., 2016. Molecular Markers for Tracking the Origin and Worldwide Distribution of Invasive Strains of *Puccinia striiformis*. *Ecology Evolution*, 6:790-2804.

Yahyaoui AH, Hakim MS, Naimi ME, et al., 2002. Evolution of Physiologic Races and Virulence of *Puccinia striiformis* on Wheat in Syria and Lebanon. *Plant Dis*., 86(5):499-504. DOI: 10.1094/PDIS.2002.86.5.499.

Zhou XL, Wang MN, Chen XM, et al., 2014. Identification of *Yr59* Conferring High-temperature Adult-plant Resistance to Yellow Rust in Wheat Germplasm PI 178759. *Theor. Appl. Genet.,* 127(4):935-45. DOI: 10.1007/s00122-014-2269-z.

Chapter 6
Wheat Yellow Rust in Egypt

Mohamed A. Gad*, Reda I. Omera and Atef A. Shahin*
Plant Pathology Research Institute, Agricultural Research Center, Giza 12619, Egypt
* Corresponding author: mohamedabo2002@yahoo.com/ atef. shahin 66 @ gmail.com

Abstract: Wheat (*Triticum aestivum* L.) is a major winter crop and an essential source of carbohydrates and multiple nutrients, serving as a dietary food in Egypt. Wheat production in Egypt is not sufficient due to various factors, including rust diseases, which are the most destructive diseases of crop plants. In Egypt, yellow rust (*Puccinia striiformis* f. sp. *tritici*) has been considered to be the most severe disease of wheat, several epidemics have been reported during the last five decades. The book chapter attempts to provide a comprehensive overview of the status and importance of wheat rust along with the research done on various aspects ranging from epidemiology to race identification, molecular genotyping, resistance screening and resistance genes utilization, and disease management. The discussion is extended in the context of regional yellow rust situation with an emphasis on potential collaboration at regional level.

Keywords: Wheat; Yellow Rust; *Puccinia striiformis* f. sp. *tritici*; Epidemics; Control; Egypt

6.1 Historical Epidemics and Impact

Wheat production across the globe was 730.55 million metric tons, while Egypt produced 8.45 million metric tons during the year 2018-2019, contributing 1.16% to world production (USDA, 2019). Wheat provides 37% of calories and 40% of protein in Egyptian food (Mujeeb et al., 2008). Therefore, the safety of wheat production plays a crucial role for food safety in Egypt. Wheat yellow rust caused by *Puccinia striiformis* f. sp. *tritici* (*Pst*) has been considered to be the most severe disease of wheat, which resulted in huge yield losses when serious epidemic occurred (Gad et al., 2019a). The yield loss due to yellow rust varied among genotypes and locations with an overall range from 12.7%~87.0% (El-Daoudi et al., 1996; Bolat and Altay, 2007), normally yield loss was approximately 20%~30% (Ashmawy and Rageb, 2016; Draz et al., 2018; Gad et al., 2020a). Under some circumstances, the disease can cause 100% yield loss when infection occurs very early and the disease continues to develop during the growing season (Afzal et al., 2007).

Egypt is located in north-eastern Africa, which has water boundaries over the Mediterranean Sea and the Red Sea, considered to be a part of the Middle East, the origin of common wheat, also the most likely source of newly spreading, high-temperature-adapted strains of *Pst* (Ali et al., 2014). Egypt belongs to the epidemiological zone of wheat rusts diseases (Saari and Prescott, 1985), where yellow rust is the most destructive disease in many wheat growing areas (Abu Aly et al., 2017). Northern governorates considered to be yellow rust hot spot in Egypt (Shahin, 2017).

Historically, wheat yellow rust was considered a sporadic disease in Egypt, while during the last five decades, several yellow rust epidemics have been occurred at different intensities in 1967, 1983, 1986, 1995 and 1997, while more recently in 2015. These epidemics severely attacked the widely grown and high yielding wheat cultivars Giza 144, Giza 150, Giza 163, Gemmeiza 1, Gemmeiza 7, Sakha 69, Sakha 93, and the long spikes Sids cultivars. Also, these epidemics were the main cause for abandoning the above wheat cultivars from agriculture and commercial production in Egypt (El-Daoudi et al., 1996; Abu El-Naga et al., 2001; El-Naggar et al., 2016; Ashmawy and Ragab, 2016). Severe yellow rust epidemic was recorded in 2018-2019 (Shahin et al., 2019), it was firstly detected on the two wheat cultivars: Sids 12 and Gemmeiza 11, which were widely grown in the Northern governorates.

The disease could severely attack most of the current wheat varieties due to their susceptibility, especially when environmental conditions are favorable for rust incidence and development (Mundt et al., 1995; Ashmawy et al., 2012; Ashamawy and Regab, 2016; El-basyoni et al., 2019; El-Orabey et al., 2019a).

6.2 Races Identification of *Pst*

Virulence surveys of the *Pst* population provide useful information for the deployment of resistance genes and guiding breeding programmes. The virulence pattern of the yellow rust pathogen is differentiated based on its reaction on a set of 17 World/European differential genotypes carrying *Yr* genes, and named by assigning a decanery value to each of the World and European differentials (Table 6.1). Races are named as two numbers separated by "E" for European, and the number is determined by Σ X, where X is the decanery value of a differential with a susceptible reaction.

Table 6.1 Differentials used to identify races of *Pst* in Egypt

No.	World Differential Set	Resistance gene(s)	Type	European Differential Set	Resistance gene(s)	Type	Decanery value
1	Chinese 166	*Yr1*	Winter	Hybrid 46	*Yr4b, YrH46*	Winter	2^0=1
2	Lee	*Yr7, Yr22, Yr23*	Spring	Reichersberg 42	*Yr7, Yr25*	Winter	2^1=2
3	Heines Kolben	*Yr2, Yr6*	Spring	Heines Peko	*Yr2, Yr6, Yr25*	Spring	2^2=4
4	Vilmorin 23	*Yr4a, YrV23*	Winter	Nord Desprez	*Yr3a, Yr4a*	Winter	2^3=8
5	Moro	*Yr10, YrMor*	Winter	Compair	*Yr8, Yr19*	Spring	2^4=16
6	Strubes Dickkopf	*YrSD, Yr25*	Winter	Carstens V	*Yr32, YrCV2, YrCV3*	Winter	2^5=32
7	Suwon 92 / Omar	*Yr4, YrSu*	Winter	Spaldings Prolific	*YrSP*	Winter	2^6=64
8	Clement	*Yr9, YrCle*	Winter	Heines VII	*Yr2, YrHVII*	Winter	2^7=128
9	*Triticum spelta* album	*Yr5*	Spring	—	—	—	2^8=256

Virulence studies revealed 114 physiological races during the period 1999-2011 in Egypt (Ashmawy et al., 2012). Youssef et al. (2003) reported that the three yellow rust races 0E0, OE64, 230E150 were the most frequent races duing 2001-2002 growing season and Ashmawy (2005) identified race 0E0, 4E0, 2E16 and 70E40 as the predominant races between 2003 and 2005. Nazim et al. (2010) reported that the two yellow rust races 0E0 and 102E22, were the most frequent during the two growing seasons of 2005-2006 and 2006-2007 (Table 6.2).

Race 0E0, 6E0, 2E0, 2E16, 4E0, 4E4, 6E5, 6E20, 18E16, 34E16, 34E20, 38E20, 70E4 were identified in 2012-2013 and race 0E0, 2E0, 2E8, 4E0, 6E16, 70E20 and 128E28 were identified in 2013-2014 (Shahin et al., 2015). Races 0E0 and 6E16 were the most frequent (each with 12.54% frequency) during 2013-2014. During 2014-2015 growing season, Shahin et al. (2019) identified a total of 10 races during 2017-2019 wheat growing seasons. The most common races during 2017-2018 were 64E0, 66E0 and 70E20 with a frequency of 26.92%, 23.07% and 23.07%, respectively. The races 450E254, 206E174, 192E192, 136E54 and 70E182 were the most common during 2018-2019 with a frequency of 22.50%, 17.5%, 15.0%, 10.0% and 10.0%, respectively. The three races 64E0, 66E0 and 70E22 were the widely distributed geographically, as each was detected in all governorates during the first season, while the 136E54, 192E192 and 206E174 during the second season (Table 6.2).

Wheat yellow rust population has been dominated by a few pathogen races each year, these predominant races did not reappear year after year in pathogen population under the Egyptian conditions, which means the *Pst* population virulence is variable over the years in Egypt. It could be explained by the frequent exchange of the lineages in this region, i.e., the

Middle East-Mediterranean region (Ali et al., 2014; Walter et al., 2016). There is no evidence of recombination while a reduced sexual reproduction capacity was reported for the Egyptian population (Ali et al., 2010).

Table 6.2 Races of *Pst* identified in Egypt since 21th century

Year	Races identified	Predominant races	Rank No. 1 race and its frequency	Reference
2001-2002	0E0, 0E64, 2E0, 2E128, 4E2, 6E134, 132E2, 142E182, 166E146, 166E148, 230E150	0E0, 0E64, 166E148, 230E150	0E0 (31.4%)	Youssef et al., 2003
2003-2005	-	0E0, 4E0, 2E16, 70E40	-	Ashmawy, 2005
2006-2007	-	0E0, 102E22	-	Nazim et al., 2010
2008-2009	0E0, 4E0, 4E16, 32E0, 108E22, 108E128, 142E20, 230E158, 238E182, 494E158	0E0, 108E22, 238E182	0E0 (12%)	Shahin and Abu El-Naga, 2011
2013	0E0, 2E0, 2E16, 4E0, 4E4, 6E0, 6E4, 6E20, 18E16, 34E16, 34E20, 38E20, 70E4	2E0, 4E0, 6E4, 70E4, 38E20	6E4 (16.66%)	Shahin et al., 2015
2014	0E0, 2E0, 2E8, 4E0, 6E16, 70E20, 128E28	0E0, 6E16	0E0 (12.54%), 6E16 (12.54%)	Shahin et al., 2015
2015	6E4, 70E20, 128E28, 150E190, 150E244, 151E244, 250E190, 450E214	0E0, 4E8, 198E154	0E0 (12.16%)	Shahin et al., 2015
2018	0E0, 0E16, 4E130, 64E0, 66E0, 70E20, 70E32, 70E182, 70E214	64E0, 66E0, 70E2	64E0 (26.92%)	Shahin et al., 2019
2019	0E0, 0E16, 4E130, 64E8, 70E182, 206E174, 450E254	450E254, 206E174, 192E192, 136E54, 70E182	450E254 (22.5%)	Shahin et al., 2019
2020-2021	68E48, 72E8, 135E16, 151E80, 160E173, 224E191, 238E143	224E191, 238E143	224E191 (12%)	Unpublished

Note: "—" means no data available.

The similarity matrix and dendrogram based on virulence data showed considerable diversity among *Pst* races either old (0E0, 2E16, 2E0, 4E0, 6E4 and 70E4) or recently emerged (0E16, 4E130, 64E0 and 66E0) in Egypt. Virulence relatedness showed clustering with respect to virulence for *Yr* genes. Accordingly, races having no virulence (0E0) or few virulence frequencies (0E16 virulent to *Yr8, Yr19*) were individually divergent. High similarities were present between races having more virulence frequency, that is, 70E4 (V *Yr2, Yr4, Yr6, Yr7, Yr22, Yr23, Yr25, YrSu*) and 6E4 (V *Yr2, Yr6, Yr7, Yr22, Yr23, Yr25*) with 85% similarity; 2E16 (V *Yr7, Yr8, Yr19, Yr22, Yr23*) and 2E0 (V *Yr7, Yr22, Yr23)* with 67% similarity. These studies, however, reveal limited information about pathogen diversity because virulence is governed by a small fraction of genome. The evolution of *Pst* races in

Egypt during 2016-2018 has been reported by Draz (2019a), who suggested the common ancestry of the current *Pst* population in Egypt belong to three races, including those of the *PstS2* lineage (Walter et al., 2016). The race identication and cultivars resistance screening procedure showed in Fig. 6.1.

Fig. 6.1 Race identification and cultivars resistance screening procedure

6.3 Molecular Characterization of the Yellow Rust Pathogen

Limited work on the application of molecular markers have been done in Egypt to understand its genetic diversity and population genetic structure.

Draz (2019a) identified molecular variation among ten *Pst* races during 2016-2018 from some governorates in Egypt among the six old (0E0, 2E16, 2E0, 4E0, 6E4 and 70E4) and four new (0E16, 4E130, 64E0 and 66E0) races, by using three SSR markers, RJ-17, RJ-18 and RJ-20. The DNA banding patterns revealed that the three SSR markers distinguished molecular polymorphisms between the old and the new *Pst* races. The number of amplified alleles varied with different primers tested. Marker RJ-20 amplified the maximum three alleles, followed by two alleles by each of RJ-17 and RJ-18. The cluster pattern constituted two distinguished main groups based on the genetic similarities using the simple matching coefficient. The first group represented the old races (0E0, 2E16, 2E0, 4E0, 6E4 and 70E4), while the new races (0E16,

4E130, 64E0 and 66E0) occupied the second one with 20% genetic similarity between them. The old races' group was separated at 60% mean similarity and formed two sub-clusters. Only one race 0E0 occupied the first sub-cluster while the second one consisted of the remaining old races, 2E0 and 4E0 (92% similarity); 6E4 and 70E4 (94% similarity) and 2E16 individually diverged at 80% clustering. In the second group, each of two new races joined together in sub-cluster, 66E0 with 64E0 (92% similarity) and 4E130 with 0E16 (94% similarity). Although the SSR markers revealed the low polymorphism feature in the races of *Pst*, it was possible to distinguish the old Mediterranean race 6E16 (PstS3; Ali et al., 2017) in Egypt from the worldwide invasive PstS2 with RJ-17 and RJ-20 (Ali et al., 2014, 2017).

El-Orabey et al. (2019b) used six SSR primers for molecular analysis of rust populations collected from Egypt and molecular marker data showed that all the tested pathotypes were genetically different and produced an unique SSR profile. Study on genetic diversity of the pathogen can provide information on its behavior in the field in terms of virulence and aggressiveness, which would also be useful for breeders. However, a more comprehensive study is required to genotype the pathogen population using comparable techniques so that the population could be compared with the regional and worldwide dataset and thus identifying the prevalence of worldwide spread lineages (Ali et al., 2017) in Egypt.

6.4 Status of Yellow Rust Resistance in Wheat Cultivars

Conventionally, breeders select wheat lines with strong all-stage resistance, which controlled by major gene and easily to be incorporated into new cultivars. Nevertheless, all-stage resistance can be easily overcome by new virulent races. Host-genetic resistance or growing wheat cultivars having sustainable resistance is still the most effective, economic and environment friendly control method, not only to minimize crop losses but also to avoid the sudden occurrence of severe epidemics in the future (Singh and Naraynan, 2000).

Under agroclimatic conditions of Gharbia governorate, Egypt, the wheat genotypes had different reaction and severity levels for last five years against yellow rust disease during wheat growing seasons of 2016-2020. Cultivars Gemmeiza 10, Giza 171 and Sakha 95 showed the lowest rust severity ranged from TRMS to 10%, while disease severity was high in the most susceptible varieties, i.e., Gemmeiza 11, and Sids 12 (Fig. 6.2). Abu El-Naga et al. (2001) reported that wheat cultivars Giza 168, Sakha 61, Sakha 93, Gemmeiza 7 and Gemmeiza 9 have satisfactory and adequate levels of yellow rust resistance during the two years of their study. Aktas et al. (2012a) reported that during the 2009-2010 seasons, Gemmeiza 11, Misr 1, Misr 2 and Sids 12 were the most susceptible genotypes and Giza 168 was moderately resistant.

Among 20 cultivars tested, five cultivars showed resistance reaction, i.e., Misr 1, Gemmeiza 10, Gemmeiza 11, Sids 13 and Sohag 3, while the remained tested cultivars showed susceptible reaction at seedling stage. While at adult plant stage, five Egyptian cultivars showed little or no infection, which was then, selected as immune or resistant cultivars, i.e., Sakha 61, Gemmeiza 10, Benisweif 4, Benisweif 5, and Sohag 3. The rest of the cultivars showed moderately resistance to moderately susceptible and susceptible reaction. The cultivars Giza 168, Sakha 94, and Gemmeiza 7 had susceptible reaction at seedling stage while it was moderately resistant to moderately susceptible at adult stage (Shahin, 2014).

Recently, El-Naggar et al. (2016) reported that the two newly released wheat cultivars: Gemmeiza 11 and Sids 12, showed susceptibility to yellow rust under field conditions in Egypt, which was widely distributed nationwide. Shahin (2017) reported that Misr 3 and Sakha 95 were resistant, probably due to uncharacterized genes. They found that some of the most important commercial genotypes, i.e., Sids 12, Giza 168, Misr 2 and Sakha 61, known as resistant to the previously characterized *Pst* races in Egypt became susceptible. Abu Aly et al. (2017) screened 12 promising wheat lines with high resistance potentiality as new sources from 57 wheat genotypes. And thought that final rust severity (FRS) and average coefficient of infection (ACI) are more appropriate indicators than area under disease progress curve (AUDPC) and rAUDPC for screening large numbers of breeding materials, as these are easily to be handled for the breeders to facilitate success of selection.

Three epidemiological parameters: FRS, AUDPC and rate of disease increase (r-value) were used to characterize Partial Resistance (PR) to *Pst* of 12 Egyptian wheat cultivars during 2015-2016 and 2016-2017. These parameters found to be lower in the PR cultivars: Sakha 94, Sakha 95, Misr 1 and Misr 2, while higher in the highly susceptible or fast-rusting cultivars: Sids 12 and Gemmeiza 11, as well as the check variety, Morocco. Additionally, higher amounts of both 1,000 kernel weight and grain yield were obtained from the PR cultivars (Abu Aly et al., 2018).

Shahin et al. (2018) reported that, high levels of resistance to the tested *Pst* isolates were displayed by the wheat cultivars, i.e., Sakha 94, Sakha 95, Gemmeiza 12, Misr 2, Shandaweel 1, Giza 171 and Misr 3, in 2017-2018. While, out of the previous cultivars, only three cv. Sakha 95, Gemmeiza 12, and Misr 3, showed high levels of resistance in 2018-2019. On the other hand, virulence to wheat cv. Sids 12, Sakha 69, Gemmeiza 9, Gemmeiza 10, Gemmeiza 11, Sakha 93, Misr 1 and Misr 2, was lower at more than 90% frequency in the two years. The highest susceptibility was recorded with Sids 12 and Gemmeiza 11 (90S), followed by Misr 1 (80S), Misr 2 (70S), Giza 163 (60S), *Yr7* and *Yr9* (80S). Cultivars, Misr 1, Misr 2, Gemmeiza 5, Gemmeiza 12, Giza 167, and Shandweel 1, exhibiting MR in the first season, lost their resistance in the second one. Cultivars such as Sids 13, Gemmeiza 7, Gemmeiza 9, Gemmeiza 10, Gemmeiza 11, Gemmeiza 12, and Shandweel 1, carrying *Yr9* gene, became susceptible and should be restricted to use alone in wheat production (Shahin et al., 2020). The commonly cultivated Egyptian cultivars, that is, Giza 168, Giza 171, Gemmeiza 9 and Gemmeiza 11, are lack of the resistance gene *Yr10* (Elkot et al., 2016). New pathogen races appeared in the last few years and caused a broke down in the resistant genotypes in Egypt. Using new techniques to develop new genotypes with durable resistance is highly required. Recently, genome-wide association study (GWAS) has been used to analyze the resistance of cultivars, five SNP markers, located on chromosomes 2A and 4A, were found to be significantly associated with the resistance. The high variation among the tested genotypes in their resistance to the Egyptian yellow rust race confirming the possible improvement of yellow rust resistance in the Egyptian wheat genotypes. The identified five SNP markers could be used in marker-assisted selection and facilitate yellow rust resistance in Egypt (Abou-Zeid and Mourad, 2021).

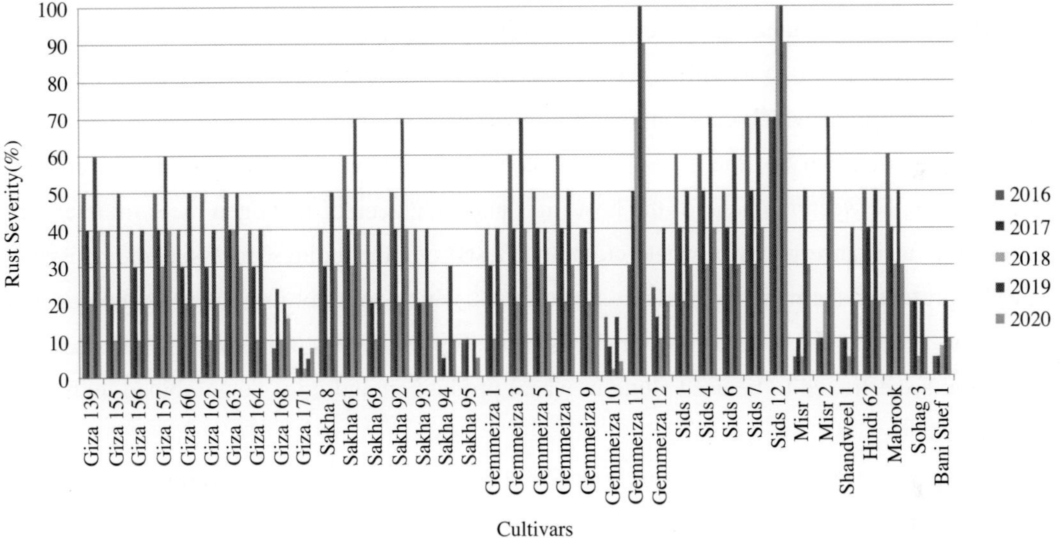

Fig. 6.2 Rust severity of different wheat cultivars in Gharbia, Egypt from 2016 to 2020

6.5 Effective and Non-effective Resistant Genes to Yellow Rust

It is a challenge to avoid yellow rust epidemics in Egypt due to lack of genetic information regarding numerous wheat cultivars, which have been released and cultivated nationwide. Those cultivars may consist of the same gene or combined genes for resistance which may lead to pressure the selection for corresponding virulent races (Ali et al., 2018). Therefore, characterizing the resistance genes in wheat cultivars is essential for developing resistant cultivars. To date, 80 resistance genes of Yellow Rust (*Yr*) have been named in wheat (McIntosh et al., 2017), out of them, 67 *Yr* genes have been temporarily designated, including seedling resistance (all-stage resistance) and adult plant resistance (APR). Among these, *Yr11, Yr12, Yr13, Yr14, Yr16, Yr18, Yr29, Yr30, Yr34, Yr36, Yr39, Yr46, Yr48, Yr52* and *Yr67* confer APR, whereas the others confer all-stage resistance (race-specific resistance), e.g., *Yr2, Yr4, Yr5, Yr6, Yr7, Yr8, Yr9, Yr10, Yr15, Yr17, Yr19, Yr25, Yr26, Yr27, Yr28, Yr35, Yr36, Yr37, Yr38, Yr40, Yr42 Yr53, Yr61, Yr65* and *Yr69* (Chen, 2005; Zheng et al., 2017). Of these, only *Yr5* and *Yr15, Yr53, Yr61, Yr65,* and *Yr69* still confer resistance to most *Pst* races and can be used in breeding for disease resistance (Shahin, 2017). Even though these *Yr* genes have been detected in various wheat varieties, but their efficacy against the diverse *Pst* pathotypes is limited due to the race specificity of all-stage resistance. By contrast, APR is generally considered durable, but it is represented by a minority among the known genes represent a minority (Chen, 2005; Ellis et al., 2014; Kankwatsa et al., 2017).

Screening of resistance genes through multipathotypic inoculation test (known as gene postulation) with a set of avirulent (*Av*)/ virulent (*v*) races of the pathogen is an effective approach to determine which genes are existed in wheat cultivars (Statler, 1984), which has been used since the discovery of the gene-for-gene concept by Flor (1959). The approach relies on the interaction between the gene of the host lines and *Av/v* gene of pathogen races to determine the probable resistance genes in wheat cultivars. Based on our findings, race-specific resistance gene *Yr9* present in the majority of Egyptian cultivars, which does not provide protection against yellow rust and may be responsible for the high susceptibility of cultivars. Cultivars carrying *Yr9* should be restricted to be used alone in wheat production and must be combined with other genes.

During wheat growing seasons of 2016-2020, results showed that the yellow rust severity for the examined monogenic lines varied from 0 to 70% with different Infection Type (IT) under natural conditions of Gharbia, Egypt. Out of 19 tested genotypes, 5 genotypes *Yr1, Yr5, Yr10, Yr15,* and *YrSP* showed desirable resistance to yellow rust disease, rust severities were 0. On the other hand, 13 genotypes showed different ITs (MR, MS, and S) with varying levels of disease severity ranged from 3%~70% (Fig. 6.3).

In Egypt, based on the reaction of the isogenic lines, the yellow rust populations in the four seasons (2013-2017) were found virulent to *YrA*, *Yr2*, *Yr6*, *Yr7*, *Yr8*, *Yr9*, *Yr17*, and *Yr27*. Avirulent to *Yr2+*, *Yr5*, *Yr10*, *Yr15*, and *Yr26*. Virulence to *Yr1* was detected only in 2015-2016, it was absent at three other growing seasons. Partial virulence was recorded in four seasons for *YrA*, *Yr18* (Anza). The Egyptian wheat genotypes had more variation for their reaction through the four growing seasons. For durable resistance in wheat, breeding programs in Egypt should focus on the use of genes like *Yr18*, *Yr29*, *Yr30*, *Yr36* and *Yr39* and many other QTLs for APR or high-temperature adult-plant (HTAP) resistance (Chen, 2005).

Yr1, Yr5, Yr10, Yr15, Yr32 and *YrSP* were resistant to yellow rust in both seedling and adult stages during the two seasons (2012-2014). While *YrA, YrSK* and *Yr18+* were susceptible at seedling stage while resistant at adult plant stage. On the other hand, *Yr6, Yr7, Yr8* and *Yr9* were susceptible at both seedling and adult stages. Wheat cultivars, i.e., Misr 1, Misr 2, Gemmeiza 10 and Giza 168 showed APR during the two seasons. Gemmeiza 11, Sids 12 and Sakha 93 were susceptible at both seedling and adult stages (Shahin et al., 2015).

The obtained results showed that *Yr1, Yr5,* and *YrSP* were the most effective during growing seasons. On the other hand, *Yr6,* and *Yr7* were attacked by a high number of races. Regarding evaluation of certain yellow rust wheat monogenic lines and Egyptian wheat varieties under the stress of both greenhouse and field conditions, the obtained results indicated that *Yr1, Yr5, Yr10, Yr15, Yr17, Yr32* and *YrSP* were resistant at seedling and adult stages. Genes such as *YrA* and *Yr18* were resistant only at adult plant stage. While testing the released wheat cultivars under natural conditions, Sakha 93 and Sids 12 were found infected (Shahin et al., 2015).

Afshari (2004) and Singh et al. (2004) reported that a new yellow rust race can attack wheat cultivars which possessing *Yr27* resistance gene in India, Yemen, Egypt, Ethiopia, Eritrea, Tajikistan, Uzbekistan and Kyrgyzstan during previous years. Thus, it is of great importance to develop wheat cultivars possessing new resistance genes for yellow rust.

Shahin et al. (2018) reported the presence of adult plant resistance (APR) gene *Yr18* in Egyptian cultivar Sakha 94. Therefore, *Yr9* gene present in Sakha 94 may be suppressed due to *Yr18* gene that conferred partial resistance to the cultivar during both seasons. Two Egyptian wheat cultivars, i.e., Giza 168 (resistant) and Giza 160 (susceptible) to yellow rust were crossed to four monogenic lines (*Yr*'s), i.e., Kalyansona, Lee, Compair, and Jupateco R, which having yellow rust resistance *Yr* gene(s); *Yr2*, *Yr7*, *Yr8* and *Yr18+*, respectively. Dominance of yellow rust resistance over susceptibility was noticed in most cases (in four out of five resistants by susceptible crosses). The *Yr8* gene had high effectiveness of conferring resistance against *Pst* under this investigation. Meanwhile, broad sense heritability estimates were high, thus early generation selection for yellow rust resistance could be effective for wheat improvement for this character (Shahin and Ragab, 2015).

Shahin (2014) suggested that accumulating 4 or 5 resistance genes to susceptible cultivars is useful to gain durable resistance cultivars. A significant step toward a better yellow rust control is identification of APR genes (Hussain et al., 1999). Recently, Hussain et al. (2011) reported that, additive, dominance and epistasis were involved in the expression of genes for yellow rust resistance in wheat. They also found that lower estimates of narrow sense heritability.

No virulence was occurred to wheat yellow rust monogenic lines with *Yr1, Yr5, Yr10* and *Yr15*, during the two years of 2017-2019 (Shahin et al., 2019). Gad et al. (2019b) recorded that the resistance genes *Yr5, Yr15, Yr17, YrTr1, Yr (7, 25) were* effective against yellow rust, during 2015-2016, 2016-2017, 2017-2018 growing seasons in Egypt. These genes can be used effectively in the breeding program for releasing new commercial cultivars under different agroclimatic conditions.

Seventeen yellow rust monogenic lines, three commercial wheat cultivars; Sakha 95, Misr 3 and Gemmeiza 11 and the highly susceptible variety Morocco were evaluated for their adult plant resistance and stability of resistance to rust diseases under different field conditions at Sadat City and Elbostan for three successive growing seasons, i.e., 2016-2017, 2017-2018 and 2018-2019. There were wide variations between genotypes. The wheat genotypes under study were classed into three classes based on the infection type. The first class included the most effective genotypes which included *Yr1, Yr5, Yr10, Yr15, YrSP*, Misr 3 and Sakha 95. The second class was genotypes of differential resistance and the third class included ineffective genotypes. Stability factors during the three growing seasons at the two locations confirmed that, *Yr17* was widely adapted and stable in its resistance (El-Orabey et al., 2019a).

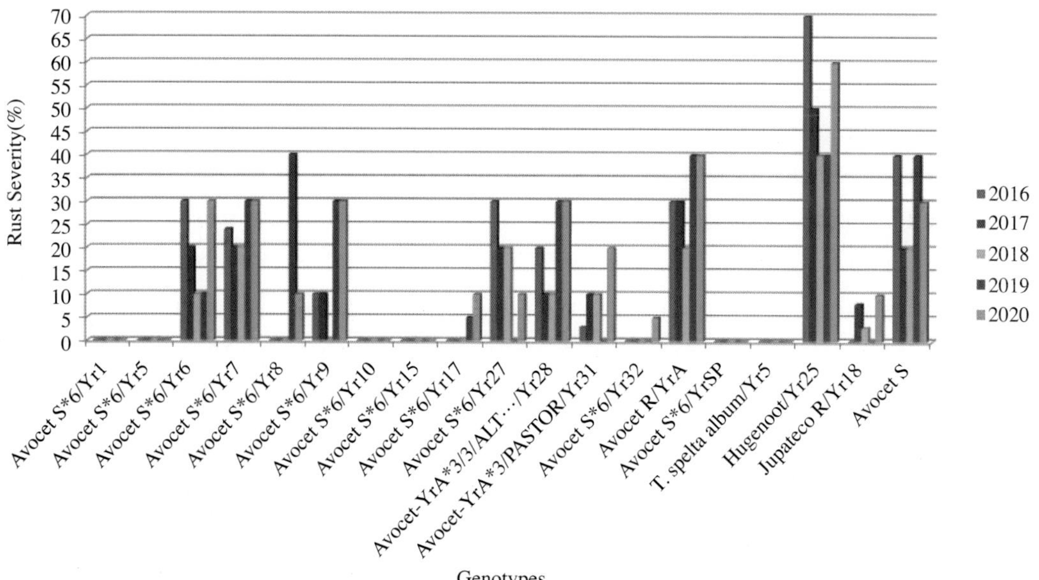

Fig. 6.3 Rust severity of the evaluated wheat lines under natural conditions of Gharbia, Egypt from 2016 to 2020

6.6 Disease Management

Genetic resistance is the most economic and effective means of reducing yield losses caused by the disease. Using resistance genes to rust disease is the most economical, effective practice, and a sustainable disease management strategy (Gad et al., 2020b). Screening of varieties against yellow rust must be a regular activity due to the dynamic evolutionary nature of the pathogen. This helps in devising and managing crops through selecting effective resistance in high yielding wheat varieties. Host materials must have a combination of effective resistance genes concerning the prevalent population of the pathogen.

Fungicides is increasingly used as a control method for stripe rust in Egypt. The widely used fungicides are listed in Table 6.3. Gad et al. (2020b) found that Tilt® 25% EC and Crwan® 25% EC had a good control effect with an over 96.91% disease control.

Table 6.3 Fungicides Used for Control Wheat Yellow Rust Disease in Egypt

Sr. No.	Fungicide	Active ingredient	Rate of application
1	Tilt® 25% EC	Propiconazole	25 ml/100 L
2	Crwan® 25% EC	Propiconazole	30 ml/100 L
3	Fungshow® 2.5% WP	Diniconazole	15 g/100 L
4	Meanar® 41% EC	Cyproconazole+Propiconazole	200 ml/100 L
5	Sumi-8® 5% EC	Diniconazole	35 ml/100 L
6	Emnent® 12.5% EW	Tetraconazole	100 ml/100 L
7	Abatchi ® 25%	Propiconazole	15 ml/100 L
8	Punch® 40% EC	Flusilazole	19 ml/100 L

Note: Cited from Gad et al. (2020b).

6.7 Future Prospective Work

Control of rust diseases has been achieved in the past largely by a combination of fungicide application along with combinations of major seedling resistance genes. However, like most race specific resistances, these major genes have lost their effectiveness, and thus there should be an increase in efforts regarding partial or adult plant resistances. Thus, efforts must be made to explore QTLs and resistance genes to pathogen populations prevalent in Egypt and other regional countries, by combining information on genetic and virulence diversity in the pathogen with genetic information on host resistance for a better rust management. For that, the acceleration of breeding program for development of yellow rust resistant genotypes should use molecular approaches, which help detection of molecular markers related to resistance or susceptible genotypes and would ensure a durable deployment of new resistance cultivar in wheat.

References

Abou-Zeid MA, Mourad AMI, 2021. Genomic Regions Associated with Yellow Rust Resistance Against the

Egyptian Race Revealed by Genome-wide Association Study. *BMC Plant Biol.*, 21(1):42. DOI: 10.1186/s12870-020-02813-6.

Abu Aly AAM, Abou-Zeid MA, Omara RI, 2018. Characterization of Partial Resistance to Yellow Rust (*Puccinia striiformis* f. sp. *tritici*) in Some Egyptian Wheat Cultivars. *J. Plant Prot. and Path. Mansoura Univ.*, 9(2): 111-119.

Abu Aly AAM, Omara RI, Abd El-Malik NI, 2017. Evaluation of New Sources of Resistance to Wheat Yellow Rust (*Puccinia striiformis* f. sp. *tritici*), under Egyptian Field Conditions. *J. Plant Prot. and Path. Mansoura Univ.*, 8(4): 181-188.

Abu El-Naga SA, Khalifa MM, Sherif S, et al., 2001. Virulence of Wheat Yellow Rust Pathotypes Identified in Egypt during 1999/2000 and Sources of Resistance. First Regional Yellow Rust Conference for Central & West Asia and North Africa 8-14 May, 2001. SPH, Karaj, Iran.

Abu-El-Naga SA, Khalifa MM, Bassiouni AA, et al., 1999. Revised Evaluation for Egyptian Wheat Germplasm against Physiologic Pathotypes of Yellow Rust. *J. Agric. Sci. Mansoura Univ.*, 24 (2):477-488.

Afshari F, 2004. Challenges of New Race of *Puccinia striiformis* f. sp. *tritici* in Iran. In: Abstracts Second Regional Yellow Rust Conference of CWANA, Islamabad, Pakistan, 22-26 March, 2004.

Afzal SN, Haque MI, Ahmedani MS, et al., 2007. Assessment of Yield Losses Caused by *Puccinia striiformis* Triggering Yellow Rust in the Most Common Wheat Varieties. *Pak. J. Bot.*, 39: 2127-2134.

Aktas A, Karaman M, Kendal E, et al., 2012a. Investigation of Yellow Rust Effect on Yield and Quality Traits of Wheat. Bursa Agriculture Fair and Congress, Publication Book.

Ali S, Gladieux P, Leconte M, et al., 2014. Origin, Migration Routes, and Worldwide Population Genetic Structure of the Wheat Yellow Rust Pathogen *Puccinia striiformis* f. sp. *tritici*. *PLoS Pathogens*, 10(1): e1003903. DOI: 10. 1371/journal. ppat. 1003903.

Ali S, Leconte M, Walker A-S et al., 2010. Reduction in the Sex Ability of Worldwide Clonal Populations of *Puccinia striiformis* f.sp. *tritici*. *Fungal Genetics and Biology*, 47:828-838.

Ali S, Sharma S, Leconte M, et al., 2018. Low Pathotype Diversity in a Recombinant *Puccinia striiformis* Population through Convergent Selection at the Eastern Part of Himalayan Centre of Diversity (Nepal). *Plant Pathology*, 67:810-820.

Ali S, Rodriguez-Algaba J, Thach T, et al., 2017. Yellow Rust Epidemics Worldwide were Caused by Pathogen Races from Divergent Genetic Lineages. *Frontiers in Plant Science*, 8:1058.

Ashmawy MA, 2005. Studies on Yellow Rust of Wheat in Egypt. Msc Thesis, Agricultural Botany Department, Faculty of Agriculture, Minufiya University, Shebin ELkom, Egypt.

Ashmawy MA, Abu Aly AA, Youseef WA, et al., 2012. Physiologic Races of Wheat Yellow Rust (*Puccinia striiformis* f. sp. *tritici*) in Egypt during 1999-2011. *Menoufia J. Agric. Res.*, 37(2):297-305.

Ashmawy MA, Ragab KE, 2016. Grain Yields of Some Wheat Genotypes to Yellow Rust in Egypt. *Menoufia J. Plant Prot.*, 1:9-18.

Ashmawy MA, Ragab KE, 2016. Grain Yields of Some Wheat Genotypes to Yellow Rust in Egypt. *Menoufia J. Plant Prot.*, 1:9-18.

Bever WM, 1937. Influence of Yellow Rust on Growth, Water Economy and Yield of Wheat and Barley. *J. Agric. Res.*, 54:375-385.

Bolat N, Altay F, 2007. Comparison of Different Methods Used in Calculating the Effect of Yellow Rust on Wheat Grain Yields. *Acta Agronomica Hungarica.*, 55(1): 89-98.

De Vallavieille-Pope C, Ali S , Leconte M, et al., 2012. Virulence Dynamics and Regional Structuring of *Puccinia striiformis* f. sp. *tritici* in France between 1984 and 2009. *Plant Dis.*, 96: 131-140.

Draz IS, 2019a. Pathotypic and Molecular Evolution of Contemporary Population of *Puccinia striiformis* f. sp. *tritici* in Egypt during 2016-2018. *J. Phytopathol.,* 167:26-34.

Draz IS, 2019b. Common Ancestry of Egyptian *Puccinia striiformis* Population along with Effective and Ineffective Resistance Genes. *Asian J. Biol. Sci.*, 12: 217-221. DOI: 10.3923/ajbs.2019.217.221.

Draz IS, Esmail SM, Abou-Zeid M, et al., 2018. Changeability in Yellow Rust Infection and Grain Yield of Wheat Associated to Climatic Conditions. *Env. Iodiv. Soil Security.*, 2:143-153. DOI: 10.21608/jenvbs.2019.6674.1040.

El-Basyoni IS, El-Orabey WM, Morsy S, et al., 2019. Evaluation of a Global Spring Wheat Panel for Yellow Rust: Resistance Loci Validation and Novel Resources Identification. *PLoS One*, 14(11): e0222755.

El-Daoudi YH, Shenoda IS, Ghaenm EH, et al., 1996. Yellow Rust Occurrence in Egypt and Assessment of Grain Yield Loss in Proc. Du Symposium Regional Sur les Maladies des Cerales et des Legumineuses Alimentaries11-14 Nov., 1996, Rabat, Maroc.

Elkot AFA, Abd El-Aziz MH, Aldrussi IA, et al., 2016. Molecular Identification of Some Stem Rust and Yellow Rust Resistance Genes in Egyptian Wheat and Some Exotic Genotypes. *Assiut Journal of Agricultural Sciences*, 47(4):124-135.

Ellis GJ, Lagudah SE, Spielmeyer W, et al., 2014. The Past, Present and Future of Breeding Rust Resistant Wheat. *Plant Science*, 5: 1-13.

El-Naggar DR, Omara RI, Abd El Malik NI, et al., 2016. Losses Assessment in Some Egyptian Wheat Cultivars Caused by Yellow Rust Pathogen (*Puccinia striiformis*). *Egypt. J. Phytopathol.*, 44(1):191-203.

El-Orabey WM, ElbasyoniI S, El-Moghazy SM, et al., 2019a. Effective and Ineffective of Some Resistance Genes to Wheat Leaf, Stem and Yellow Rust Diseases in Egypt. *J. Plant Production, Mansoura Univ.*, 10 (4): 361-371.

El-Orabey WM, Hamwieh A, Gad MA, et al., 2019b. Virulence and Molecular Polymorphism of *Puccinia triticina* pathotypes in Egypt. *Int. J. Phytopathol.*, 8 (3):111-122.

Flor HH, 1959. Genetic Controls and Host Parasite Interactions in Rust Diseases. In: Holton CS (ed.). Plant Pathology, Problems and Progress 1908-1958. Madison: University of Wisconsin Press.

Gad MA, El-Naggar DR, El-Orabey WM, et al., 2020a. Characterization of Virulence and Diversity of *Puccinia graminis* f. sp. *tritici* on Wheat in Egypt. *Egypt. JAgron.*, 42(1):19-34.

Gad MA, Khaled YA, Fayza AS, et al., 2020b. Soliman. Comparative of Fungicidal Efficacy against Yellow Rust Disease in Wheat Plants in Compatibility with Some Biochemical Alterations. *Menoufia J. Plant Prot.*, 5: 29-38.

Gad MA, Li HX, Li MJ, et al., 2019b. Evaluation of Wheat Genotypes to Rust Diseases (*Puccinia* spp.) under Agroclimatic Conditions of Egypt and China. *Journal of Agricultural and Crop Research*, 7(9):170-180.

Gebril EEMA, Gad MA, Kishk AMS, 2018. Effect of Sowing Dates on Potential Yield and Rust Resistance of Some Wheat Cultivars. *J. Plant Production, Mansoura Univ.*, 9 (4): 369-375.

Hasan MA, Boult OA, Abou-Zeid M, et al., 2016. Impact of Different Levels of Stem and Yellow Rust Severities on Two Grain Yield Components of Wheat. *Minufiya J. Agric. Res.*, 41(3).

Hussain M, Chaudhry MH, Rehman A, et al., 1999. Development of Durable Rust Resistance in Wheat. *Pakistan J. phytopathology*, 11:130-139.

Kankwatsa P, Singh D, Thomson PC, et al., 2017. Characterization and Genome-wide Association Mapping of Resistance to Leaf Rust, Stem Rust and Yellow Rust in a Geographically Diverse Collection of Spring Wheat Landraces. *Mol. Breed.*, 37:113. DOI: 10.1007/s11032-017-0707-8.

McIntosh RA, Dubcovsky J, Rogers WJ, et al., 2017. Catalogue of Gene Symbols for Wheat: 2017 Supplement. Available online: http://shigen.nig.ac.jp/wheat/komugi/genes/macgene/ supplement2017.pdf. Mundt CC, Brophy LS, Schmitt MS, 1995. Disease Severity and Yield of Pure-line Wheat Cultivars and Mixture in the Presence of Eye Spot, Yellow Rust, and their Combination. *Plant Pathology*, 44: 173-182.

Menshawy AM, Najeeb MAA, 2004. Genetical and Pathological Studies on Certain Egyptian Wheat Genotypes as Affected both Leaf and Yellow Rust. *J. Agric. Sci. Mansoura Univ*, 29(4): 2041-2051.

Nazim M, Awad MA, Khalifa SZ, et al., 2010. Fréquence of Virulence and Virulence Formula of Wheat Yellow Rust Races Identified in Egypt. *Menoufia J. Agric. Res.*, 35(2): 439-452.

Perronne R, Dubs F, de Vallavieille-Pope C, et al., 2021. Spatiotemporal Changes in Varietal Resistance to Wheat Yellow Rust in France Reveal an Increase in Field Resistance Level During the Period 1985–2018. *Phytopathology*, https://doi.org/10.1094/PHYTO-05-20-0187-R.

Saari EE, Prescott JM, 1985. World Distribution in Relation to Economic Losses. In The Cereal Rusts, vol. 2, Diseases, Distribution, Epidemiology, and Control. Roelfs AP & Bushnell WR (eds). Orlando: Academic Press.

Shahin AA, 2014. Resistance to Yellow Rust in Some Egyptian Wheat Germplasm. *J. Plant Prot. and Path. Mansoura Univ.*, 5 (11): 983-993.

Shahin AA, 2017. Effective Genes for Resistance to Wheat Yellow Rust and Virulence of *Puccinia striiformis* f. sp. *tritici* in Egypt. *Egypt. Acad. J. Biol. Sci.*, 8 (2): 1-10. DOI: 10.21608/EAJBSH.2017.16762.

Shahin AA, Abu Aly AA, Shahin SI, 2015. Virulence and Diversity of Wheat Yellow Rust Pathogen in Egypt. *Journal of American Science*, 11(6):47-52.

Shahin AA, Abu El-Naga SA, 2011. Physiological Race Diversity and Virulence of *Puccinia striiformis* at Both Seedling and Adult Plants of Wheat in Egypt. *Arabian Journal of Plant Protection*, 29(1): 90-94.

Shahin AA, Draz IS, Esmail SM, 2020. Race Specificity of Yellow Rust Resistance in Relation to Susceptibility of Egyptian Wheat Cultivars. *Egypt. J. Phyto.*, 48:1-13.

Shahin AA, Esmail SM, Abd El-Naby H, 2019. Virulence Dynamics and Diversity of *Puccinia striiformis* Populations in Egypt during 2017/2018 and 2018/2019 Growing Seasons. *J. Plant Prot. Pathol. Mansoura Univ.*, 10 (12):655-666.

Shahin AA, Omar Hend A, El-Sayed AB, 2018. Characterization of *Yr18/Lr34* Partial Resistance Gene to Yellow Rust in Some Egyptian Wheat Cultivars. *Egypt. J. Plant Prot. Res.*, 6 (3):1-9.

Shahin AA, Ragab KE, 2015. Inheritance of Adult Plant Yellow Rust Resistance in Wheat Cultivars Giza160 and Giza168. *J. Plant Prot. and Path. Mansoura Univ.*, 6 (4): 587-596.

Singh P, Naraynan SS, 2000. Biometrical Techniques in Plant Breeding. Ludhiana: Kalani Publishers.

Singh RP, William HM, Huerta-Espino J, et al., 2004. Wheat Rust in Asia: Meeting the Challenges with Old and New Technologies. In: Proceedings of the 4th International Crop Science Conference, 26 Sep. - 1 Oct., 2004. Brisbane, Australia.

Statler GD, 1984. Probable Genes for Leaf Rust Resistance in Several Hard Red Spring Wheats. *Crop Sci.*, 24:883-886. https://doi.org/10.2135/cropsci1984.0011183X002400050013x.

USDA, 2019. World Agricultural Production. Circular Series WAP 8-19 August 2019 United States Department of Agriculture, Washington DC, USA. http://www.nass.usda.gov/Publications/.

Walter S, Ali S, Kemen E, et al., 2016. Molecular Markers for Tracking the Origin and Worldwide Distribution of Invasive Strains of *Puccinia striiformis*. *Ecol. Evol.*, 6: 2790-2804.

Youssef WA, Nagib MA, Matelda F, et al., 2003. Wheat Yellow Rust Pathotypes, their Frequency and Virulence Formulae in Egypt during 2000/2001 & 2001/2002. *J. Agric. Sci. Mansoura Univ.*, 28(5): 2489-3477.

Zheng SG, Li YF, Lu L, et al., 2017. Evaluating the Contribution of *Yr* Genes to Yellow Rust Resistance Breeding through Marker Assisted Detection in Wheat. *Euphytica*, 213:50. DOI: 10. 1007/s10681-016-1828-6.

图书在版编目（CIP）数据

泛喜马拉雅地区和中东小麦条锈病 = Wheat Yellow Rust in the Extended Himalayan Regions and the Middle East ：英文 / 李明菊，（巴基）萨吉德 · 阿里著. —北京：中国农业出版社，2022.8

ISBN 978-7-109-29558-2

Ⅰ. ①泛… Ⅱ. ①李…②萨… Ⅲ. ①小麦－条锈病－研究－英文 Ⅳ. ①S435.121.4

中国版本图书馆CIP数据核字（2022）第108654号

中国农业出版社出版

地址：北京市朝阳区麦子店街18号楼

邮编：100125

责任编辑：闫保荣　　文字编辑：杨　春

版式设计：杨　婧　　责任校对：吴丽婷　　责任印制：王　宏

印刷：北京缤索印刷有限公司

版次：2022年8月第1版

印次：2022年8月北京第1次印刷

发行：新华书店北京发行所

开本：787mm × 1092mm　1/16

印张：9.25

字数：270千字

定价：98.00元
